Remote Control Robotics

Springer
New York
Berlin
Heidelberg
Barcelona
Hong Kong
London
Milan
Paris
Singapore
Tokyo

Craig Sayers

Remote Control Robotics

With 75 Illustrations, 11 in full color

Craig Sayers
csayers@acm.org

Library of Congress Cataloging-in-Publication Data
Sayers, Craig.
Remote control robotics / Craig Sayers.
p. cm.
Based on the author's Ph.D. dissertation, University of Pennsylvania.
Includes bibliographical references and index.
ISBN 0-387-98597-2 (alk. paper)
1. Robots—(Control systems). 2. Remote control. I. Title.
TJ211.35.S29 1998
620'.46—dc21 98-29989

Printed on acid-free paper.

Production managed by Anthony K. Guardiola; manufacturing supervised by Thomas King.
Camera-ready copy prepared using the author's TeX files.
Printed and bound by Maple-Vail Book Manufacturing Group, York, PA.
Printed in the United States of America.

9 8 7 6 5 4 3 2 1

ISBN 0-387-98597-2 Springer-Verlag New York Berlin Heidelberg SPIN 10687977

For Professor Paul

Abstract

Our goal is to allow you, a human operator, to control a machine, a remote robot. If you and the robot both occupy the same room, then this is a relatively well understood problem. But consider what happens if we move the remote robot to a laboratory in the next county and supply you with a connection via a constrained communications channel such as the Internet. Now there is insufficient bandwidth to provide you with a high resolution view of the robot and, even worse, there is a significant time delay. If the robot were walking and it stumbled, then it would crash to the ground before you even received the first video image showing any problem.

Now, consider what happens if we move the remote robot out of the laboratory and into the real world. Perhaps it is on an uneven footpath, or perhaps it is submerged on the sea floor. Now, not only can you not correct if something goes wrong, but it's much more likely that something will go wrong. That's the subject of this text. Our aim is to let you control a remote robot efficiently, in a real environment, via a constrained communications link.

We'll begin with an introduction to the basics of robotics, take a historical look at controlling remote machines, and then examine the difficulties imposed by delayed, low-bandwidth, communications. To overcome the problems of constrained communications we'll introduce the teleprogramming paradigm; to help a human interact more efficiently with the machine, we'll introduce active force and visual clues; to cope with the unexpected, we'll introduce techniques for diagnosing and recovering from errors; to show that the ideas are feasible, we'll describe real working implementations.

Experimental systems have been created in a laboratory, then moved to a test tank, and finally migrated to the ocean. In the most recent experiments, an operator in Philadelphia successfully performed retrieval operations using a remote manipulator on an unmanned vehicle submerged at the Massachusetts coast. All communication over the 500 km inter-site distance was via a combination of the Internet and simulated subsea acoustic modems. These tests demonstrated the feasibility of performing real-world intervention in unpredictable environments using delayed, low-bandwidth, communication links.

Acknowledgments

This manuscript is based in large part on my Ph.D. dissertation which was completed in the General Robotics and Active Sensory Perception Laboratory of the University of Pennsylvania under the guidance of Richard Paul.

I must firstly thank my advisor, Richard Paul. He provided advice, encouragement, humor, and enthusiasm in just the right amounts and at just the right times. I would also like to acknowledge the assistance of my dissertation committee: Norm Badler, Vijay Kumar, Max Mintz, and Dana Yoerger, all of whom contributed to this document.

Thanks also to Martin Gilchrist at Springer-Verlag for recognizing that there was a book hidden inside the dissertation, and to all the other folks at Springer-Verlag who helped see it through to completion. Special thanks go to the readers: Brian Beach, Sharon Beach, Richard Paul, and Brian Wilson and to the copy editor, Chrisa Hotchkiss. Any remaining mistakes are mine.

The teleprogramming paradigm came from the work of Janez Funda and Tom Lindsay under the guidance of Richard Paul. Much of my early insight into the teleprogramming system came from discussions with them, and it was their implementation that served as the prototype for the current system.

The initial in-air testing was aided by Matt Stein's work on the slave station. A subsequent in-air slave implementation was possible due to the assistance of Zahed Wahhaj and Sicong Li. The camera calibration scheme at the operator station would not have been possible without Angela Lai's careful implementation of Tsai's algorithm while the real-time visual imagery was greatly assisted by Matt Grund's implementation of the JPEG algorithms.

A number of staff members at the GRASP lab assisted in maintaining the hardware and software employed. They included Fil Fuma, John Bradley, Mark Foster, Glenn Mulvaney, and Gaylord Holder. Russ Anderson and AT&T Bell Labs donated several pieces of equipment, including the JIFFE coprocessor on which the system relies.

The subsea experiments were only possible because of the active assistance of a number of people at the Woods Hole Oceanographic Institution, especially Dana Yoerger, Louis Whitcomb, and Josko Catipovic.

The JASON subsea vehicle was managed by Andy Bowen, while Will Sellers and Bob Elder did an excellent job of handling and piloting it under what were, at times, very difficult conditions. The success of the later experiments was certainly helped by Tad Snow's elegant gripper design, while Tim Silva helped us capture the results on videotape.

A number of other people have contributed to the JASON vehicle and these experiments. They include (in alphabetical order) Bob Ballard, Tom Crook, Dick Edwards, Larry Flick, Skip Gleason, Bevan Grant, David Hoag, Bill Lange, Bob McCabe, Cindy Sullivan, Mary Jane Tucci, and Nathan Ulrich. Thanks also to Dave McDonald and the WHOI security team.

Teleprogramming System Development was funded in part by NSF Grant BCS-89-01352, ARPA Grant N00014-92-J-1647, ARO Grant DAAL03-89-C-0031PRI, and NSF Grant CISE/CDA-88-22719. The JASON ROV System is operated by the Woods Hole Oceanographic Institution's Deep Submergence Operations Group, supported by NSF Grant OCE-9300037. The Teleprogramming In-water Tests were funded by the Naval Underwater Warfare Center, Newport, RI, under contract number N66606-4033-4790.

I am especially grateful to the Woods Hole Oceanographic Institution for providing a postdoctoral scholarship, considerable academic freedom, and some wonderful memories.

Contents

List of Figures

1

Introduction

"...since the sea is our most universal symbol for memory, might there not be some mysterious affinity between these published recollections and the thunder of waves? So I put down what follows with the happy conviction that these pages will find their way into some bookshelf with a good view of a stormy coast. I can even see the room—see the straw rug, the window glass clouded with salt, and feel the house shake to the ringing of a heavy sea."

—The Stories of John Cheever, Ballantine Books, 1991.

Imagine that you are sitting in a chair, perhaps even one with a view of the sea, and that I place a baseball-sized sphere in your hand, connected to the back of the sphere is a mechanical linkage that disappears up into the ceiling. Now imagine that on the floor at your feet I place a robot arm. On the end of the robot arm is another baseball-sized sphere and somehow, as if through magic, the sphere in your hand is connected to the one on the end of the robot. Now, whenever you move your hand left, the robot moves left. Whenever you move your hand up, the robot moves up. This mode of operation, where you, an operator, directly control a robot is termed *teleoperation*, and its historical development will be described in Chapter 3.

Now, imagine that you move your hand down, causing the robot to move down. When the sphere on the end of the robot contacts the floor, you feel the sphere in your hand stop moving. Even though your hand is still above your lap, it feels as though you had reached down and touched the floor yourself. Any force you apply to the sphere in your hand is duplicated by the sphere on the robot, and any force felt by the robot is duplicated on your hand. This is called *bilateral teleoperation.*

Now, imagine that I take the robot and move it to the other side of town. Then I place a TV screen at your feet and connect it to a camera pointed at the robot. Now, you can control the robot just as before. When you move your hand, the real robot and its TV image both move. Only now your view is a little more restricted; since the TV picture is flat you can't judge depths quite so well, and if you lean forward in your chair you just see the back of the TV set, and not the back of the robot as you could before.

Now, imagine that I take the robot and the camera and move them further away. You can still see the robot on the TV. It looks slightly larger than before, though perhaps that is just your imagination, and the ground around it looks strange. Curious about what it feels like, you move to touch it. But nothing happens, so you move some more, but still nothing happens. Then suddenly, after several seconds, you see the TV image of the remote robot begin to move, and you recognize that it is doing what you did several seconds ago. Pausing to think, you realize that the connection between your sphere and its sphere is not magic at all, for if it were, then things would not become delayed as they became further removed. But your musing is interrupted, for suddenly your hand is jolted upwards. Looking at the TV, you see that the remote robot has smashed its sphere against the ground—the force you just felt was the force it felt several seconds ago. Conventional teleoperation does not work well in the presence of communication delays.

To avoid problems caused by the delay, I could replace the TV with a computer-generated display and show you, not what the robot was doing now, but instead what it would do when it tried to duplicate your motions. When you move your sphere, the simulated robot on the computer screen responds immediately. You can control it (and hence indirectly control the real remote robot) in much the same way as when the robot was right at your feet. This type of system is called a *predictive display* (see Chapter 3).

On its own, the predictive display is not sufficient, since the computer can't simulate the remote environment perfectly. Thus, to make the system work (and avoid smashing any more spheres) we need to add some local intelligence to the remote robot. Not only does that intelligence help protect the robot, but it also allows us to communicate with it using higher-level symbolic commands, thus making it feasible to communicate over links that have low bandwidth as well as high latency. This system is termed *teleprogramming* (see Chapter 5).

The computer needs a model of the remote environment in order to create the simulation. If we embellish that model, adding a little knowledge of task actions you may wish to perform, then the computer could attempt to predict and then actively assist your motions. For example, if you wanted to move in a straight line, it could create a virtual ruler to assist you. You could see that ruler on the screen, and feel the sphere in your hand slide against it, even though there is no real ruler at the real remote site. The mechanism used to achieve this is termed *synthetic fixturing*, and it is introduced in Chapter 7.

Earlier in the day, when the robot you were teleoperating was just on the other side of town, the visual feedback provided by the TV screen gave you considerable information about what was happening at the remote site. If something unexpected were to occur, you would probably have been able to see it. In the move to teleprogramming using a simulated display, that ability has been lost. Since one of the advantages of teleprogramming is its ability to operate over low-bandwidth communications channels, it is not desirable to give up that benefit in order to get back the TV images. However, if the system could predict exactly where you would have wanted to look, then it could select an appropriate fragment of the remote camera's view and send just that one portion back to you. Furthermore, if there were several cameras, it could select from among them to find the one with the best view. This may be achieved using *intelligent fragmentation*, and it is described in Chapter 8.

Later in the day, when you smashed the remote robot's sphere against the ground, it could be argued that, had you been aware of the time delay, such accidents could have been avoided. However, regardless of how well prepared the operator is, and regardless of how well engineered the system is, there will still be cases where unexpected events occur. Thus, if a practical implementation is to be developed, then we must have a mechanism for detecting, diagnosing, and recovering from, such unpredictable occurrences. The operator interface that makes this possible is described in Chapter 9.

Combining a graphical simulation, symbolic communication, some level of remote-site intelligence, and a sophisticated interface for error diagnosis and recovery should permit operators to control remote manipulators while using low-bandwidth, high-latency communications links. Such links include the Internet, long-distance space communication, and wireless subsea modems.

1.1 The fundamental tradeoff

In controlling a remote robot via constrained communications, information is a valuable resource. Having additional knowledge at one site of data stored at the other site can speed task performance, predict and avert errors, and speed recovery from those unexpected events that do occur. However, sending information between sites is expensive—it uses valuable bandwidth, takes time and, perhaps most importantly, it consumes power. Thus, there is a continual need to trade off between the cost of sending additional information between sites and the possible time saving resulting from having that additional information immediately available. The need to make this tradeoff will be revisited many times throughout this book.

1.2 Automation

Just as we must make tradeoffs when sending information, so we must also make tradeoffs when deciding how much to automate. Our goal is an appropriate level of technology.

Imagine that you are teaching children to build a treehouse. One extreme approach would be to sit down with them beforehand and describe exactly how the project should progress, listing everything that might happen, discussing what to do in each case, and then sending them out into the garden alone to begin construction. This is analogous to current attempts at autonomous robotic systems.

An extreme alternative is to watch over the children at every step holding their hands during the entire process. This makes it very likely that you'll end up with a nice treehouse but it requires your full attention and necessitates your being actively involved during the entire process. That's analogous to current remotely operated robots.

A more moderate approach is to tell the children a few steps at a time, listening and glancing out of the window occasionally to make sure everything is progressing as planned and lending a hand only when they appear to be having difficulty.

We're aiming to give human operators of remote robots an analogous feeling. It's not practical to have an operator supervise the minutest detail of operation at a remote site—the delays are just too great. But neither is it possible to completely remove the operator—even the most powerful modern robots are still incredibly stupid in comparison to a young human. Thus, we need to compromise; providing just enough autonomy at the remote site so that the operator can feel comfortable letting it work on its own for a minute.

1.3 Subsea robotics

Now, consider a more realistic situation and one to which we'll return later in the text. Imagine the task of servicing a piece of equipment on the sea floor and assume that, for reasons of safety and economics, we wish to perform this operation without requiring direct human intervention.

One technique we might attempt is a purely autonomous approach. There are existing autonomous underwater vehicles (AUVs) that can transit to a target area, perform a search, and return to a waiting vessel. Thus, equipping such a vehicle with a robot manipulator and programming it to perform a servicing task does not, at first, seem such an arduous task. Unfortunately, it will not work.

The reason is that performing a manipulative task involves a fundamentally different scale of interaction between machine and environment. Here the AUV is not just observing its environment, but is instead actively changing it. Thus, the vehicle must be concerned not just with its own internal state, but also with the changing state of the equipment on which it is working. This is complicated by those unexpected events that will inevitably occur during real-world interaction. Parts may fail or break, tools may slip or jam. If such a task is to be performed reliably then we have two choices. Either we must predict and preprogram solutions to every possible eventuality, or we must make a machine with sufficient intelligence that it can replace the knowledge and resourcefulness of experienced humans. Neither of these possibilities is currently feasible, and neither is likely to be feasible in the near future.

An alternative technique for performing the servicing operation, and the one currently employed commercially, is the use of a remotely operated vehicle (ROV). In this case, a platform containing cameras and robot manipulators is positioned near the work site and connected to a surface ship via a long wire tether. Operators aboard that ship perform the servicing task. They control the position of the platform and command the remote manipulators while observing images transmitted up the tether from the underwater cameras. This form of teleoperation has been used underwater since the early 1960s [90]. While the employed equipment has been improved, the basic strategy has remained unchanged. It relies on the continuous presence of a direct wired link between operator and subsea vehicle.

We desire a system that combines the best of both the autonomous and conventional teleoperation approaches. Just like the autonomous system, we desire a vehicle that does not require a direct wired connection to the surface. Just like the teleoperation system, we desire an implementation where the human operator may diagnose and correct problems as they occur, without any need to predict beforehand everything that may go wrong.

A solution is to allow operators ashore to communicate with the vehicle using a wireless connection. Unfortunately, the only long-range wireless underwater communications scheme is via an acoustic modem link [21]. Compared with a wired tether, such links provide only a tiny bandwidth (on the order of 10 kbits/sec) and, since sound travels at only 1450–1550 m/s in water [20], they add a relatively large communications delay (approaching 10 seconds round-trip). In this manuscript, we show that it is possible to perform real-world manipulative tasks even in such a constrained communications environment.

1.4 Chapter overview

Chapter 2 introduces readers unfamiliar with the robotics literature to the terminology employed later in the book. In particular, we'll look at degrees of freedom, redundancy, forward and inverse kinematics, singularities, and simple input/output devices.

Chapter 3 presents a brief overview of historical developments in teleoperation. While early implementations were relatively simple, their performance for some tasks is unlikely to be bettered—even by the most modern implementations. When interacting with the real world, it is the operator's ability to overcome the unexpected, and the system's ability to show him or her when the unexpected has occurred, which becomes the limiting factor on performance. This chapter will also summarize prior work on operator aids and look at moves to increase the separation between human operator and remote environment.

Chapter 4 describes teleoperation systems that are more modern, though in some sense less able. These rely on the Internet as a communications medium. They allow an operator to control a remote robot via a low-bandwidth and variable-latency communications scheme. By observing how these systems operate, readers may begin to appreciate the difficulty of remote control via constrained communications.

Chapter 5 describes the teleprogramming paradigm. This provides the means to efficiently control remote robots via constrained communications. It works by providing a low level of autonomy at the remote site, a graphical simulation at the operator station, and by using symbolic inter-site communication. It is suggested that the graphical simulation insulates the operator from the communications delay in much the same way that caches insulate modern central processing units from the time delays required for communication with dynamic memory.

Chapter 6 introduces the notion of a natural operator interface and describes the mechanism by which operator motion may be translated into the motion of a real or simulated robot manipulator. Consideration is also given to interesting cases where the operator input device has a very different structure than the remote robot. In particular, we consider the case where the operator uses a computer mouse to command an industrial robot.

Chapter 7 introduces the notion of synthetic fixturing, in which the system uses active force and visual clues to assist operators in teleprogramming task performance.

Chapter 8 introduces the notions of intelligent fragmentation, intelligent frame rate, and intelligent task rate. Combined with conventional image compression, these allow the system to provide the operator with real-time visual imagery via a very low-bandwidth communications channel.

Chapter 9 describes the difficult task of detecting, diagnosing, and recovering from, those unexpected events that will inevitably occur during interaction with the real world.

Chapter 10 describes a mechanism by which operator actions are observed and translated into symbolic commands for transmission to the remote robot.

Chapter 11 describes working implementations of the teleprogramming system that combine a graphical simulation, synthetic fixturing, symbolic communication, intelligent fragmentation, remote-site autonomy, and real-time visual imagery. Early systems operated in the laboratory with only a few feet between sites, later implementations moved the slave robot into a test tank, and then finally into the ocean. In recent trials, an operator in Philadelphia was successful in controlling a robot manipulator mounted on the JASON subsea vehicle and submerged on the sea floor at Woods Hole, Massachusetts. The required communication bandwidth was less than that available from a household telephone line, while the communications delay was in excess of 10 seconds.

Chapter 12 presents the experience gained building systems for interacting with a very real and unpredictable world. It extends concepts from our implementation, providing ideas for future research.

Chapter 13 concludes the manuscript with a brief summary and predictions for the future.

2

Basics

In this chapter we'll define the terminology and, with the aid of a few toy examples, introduce some basic concepts for use later in the text. Those readers who have taken a college-level course in robotics will probably want to skip this chapter.

For the sake of brevity, precise mathematical definitions have been sacrificed in favor of short intuitive descriptions. For a more detailed and lengthy treatment, readers are referred to any of several excellent robotics books. (See, for example, the classic text by Paul [59], or more recent works by Craig [18] and Yoshikawa [98].)

2.1 Single link robot

Consider first the case of a robot manipulator with just one rotating or *revolute* joint. Being capable of motion about only a single axis, this robot is considered to have one *degree of freedom (DOF)* (see Figure 2.1).

Imagine that we have two such robots. One is on a coffee table in front of you, while the other is holding a satellite antenna on the roof of your house. By moving the small manipulator on your coffee table, you can control the orientation of the robot antenna on the roof. Now, if we are to allow you to

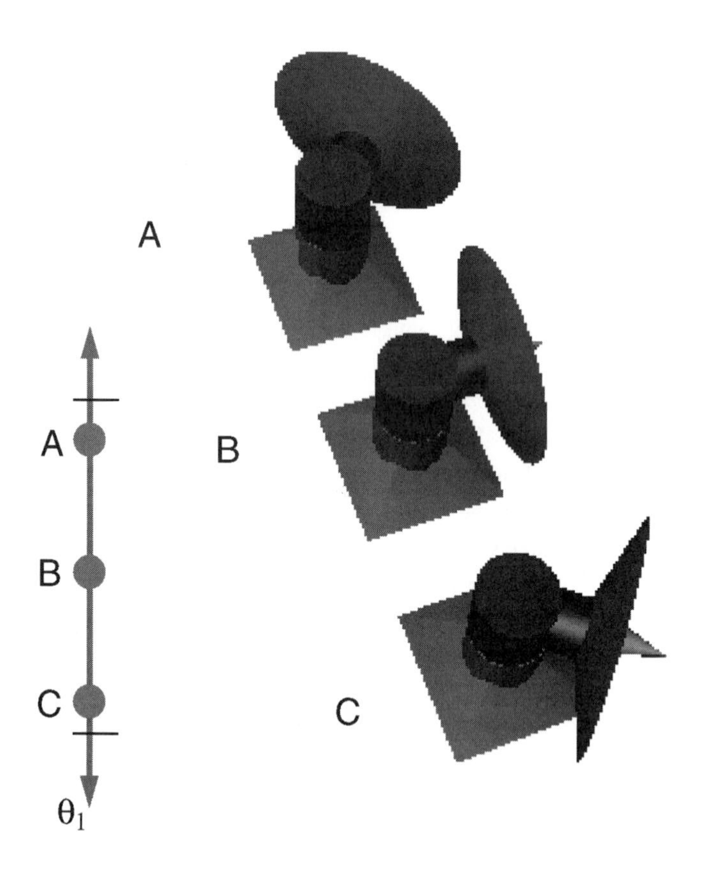

FIGURE 2.1: This satellite antenna is capable of motion about only a single axis. In some sense, it may be considered a one-degree-of-freedom robot. The joint space representation for this system is a single line.

not just move the roof antenna, but instead program it to move at particular times, then we need the means to record the position of the robot. To do that, we need a way to represent its position symbolically. The symbolic representation for manipulator position falls under the general area of *robot kinematics.*

Consider the robot on the roof. Its position may be uniquely defined as the amount by which the joint has rotated. Imagine a line whose starting point is the minimum angle to which the joint may rotate and whose ending point is the maximum angle to which the joint may turn. That line is a *joint space* representation for this simple one DOF manipulator. Any particular position of the robot may be described by a point on that joint space line. Motion of the robot's joint is equivalent to sliding along that line.

Assuming that the robot on your coffee table is an exact scaled copy of the robot on your roof (i.e. that they are *kinematically similar*), then we may record your motion of the table-top manipulator by storing the corresponding points in joint space. Those may then be used to directly command the motion of the roof-top robot.

2.2 Two-link toy

Now consider another example, this time with two degrees of freedom—a child's Etch-a-Sketch™[1] toy. This marvelous device has a screen, behind which is a pointer and beside which are two control knobs. One knob makes the pointer move up and down, while the other makes it move left and right. By appropriately manipulating the knobs, an operator may drag the pointer around to draw a picture on-screen (see Figure 2.2).

In this case, the operator input device is the two knobs, while the output device is the pointer behind the screen. Since there is a direct mechanical connection between each knob and each axis of motion, we can describe it as a bilateral system. The practical consequence of this is that when the pointer hits a mechanical stop at the edge of the screen, you can feel that resistance on the appropriate control knob. If someone were to reach behind the screen and move the pointer, the operator's control knobs would also move.

[1] Etch-a-Sketch is a registered trademark of the Ohio Art Company.

Just as for the robot on your roof, this toy's position may be uniquely described by a point in joint space. However, since there are now two joints, the joint space representation will be a two-dimensional plane. Since the motion of each joint is limited, the region of that plane within which the robot may exist will form a rectangle. Each point within that rectangle corresponds to one unique position, while each side corresponds to a case where one of the joints is at one motion limit.

To more formally define the on-screen location of the pointer we could define a Cartesian x-axis to be along the bottom of the screen, and a y-axis to be along the side. Now the position of the pointer may be uniquely described either as a point in joint space (d_1, d_2) or, equivalently, as the pointer position in Cartesian coordinates (x, y).

The use of sliding or *prismatic* actuators creates a convenient linear relationship between joint space and Cartesian space. For each position of the joints there is one, and only one, corresponding point in Cartesian space. When you use the toy, it makes no difference whether you think of the control knobs as controlling the joints of the manipulator or if you think of them as controlling the Cartesian position of the pointer on-screen. Both are equivalent.

2.3 Two-link robot

Now, imagine a device that performs a similar function to the Etch-a-Sketch™ toy. It still has two degrees of freedom, still has two control knobs, only now the actuators use revolute joints connected in a serial linkage (see Figure 2.3). This linkage starts at the robot's base, continues through the first rotating joint, along the first link, up through the second joint, and along the second link before finally reaching the tip, or *end-effector*.

The joint space representation for this two-degree-of-freedom manipulator is almost identical to that for the Etch-a-Sketch™ toy. The only minor difference is that now the axes are marked off in degrees or radians, rather than inches or millimeters.

To allow you to command the position of this robot, we could allow you control over the location of the corresponding point in joint space. Imagine that one of the knobs controlled the horizontal position of the point in joint space, while the other controlled the vertical position. By appropriately manipulating those two controls, you could make the joint space

FIGURE 2.2: This Etch-a-Sketch™toy serves as a simple example of a two-degree-of-freedom bilateral system employing prismatic actuators.

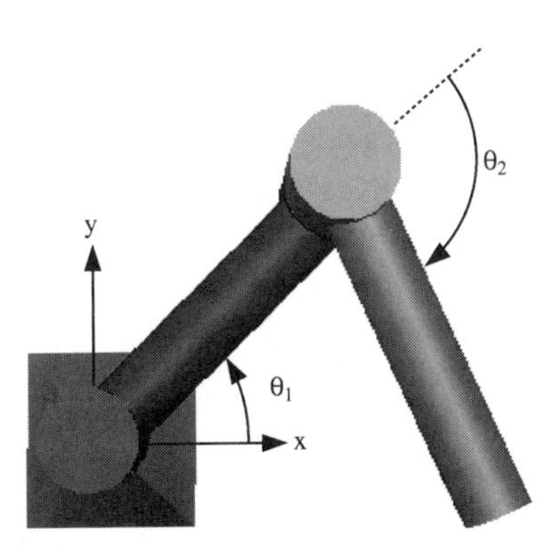

FIGURE 2.3: A two-degree-of-freedom manipulator with revolute joints.

point follow any trajectory and hence make the robot perform any feasible motion. This is called *joint space control.*

Controlling the robot in this way is very simple to implement. Each of your control knobs may be directly connected to one joint on the manipulator. However, commanding the manipulator end-effector to a particular Cartesian location can be awkward. With the Etch-a-Sketch™ there was a natural mapping between the joint space and Cartesian space representations. To move the end-effector along a straight line in Cartesian space, you had only to move the joints along a straight line in joint space(see Figure 2.4). With the switch to revolute joints, that natural mapping is missing (see Figure 2.5). Now moving along a straight line in Cartesian space requires you to trace a curve in joint space.

To allow you to control the manipulator in a more natural manner, we may desire a *Cartesian space control* system where each of your two control knobs controls the position of the end-effector in a different Cartesian axis. One knob moves the tip horizontally, while the other moves it vertically. Given such a control implementation, you could conveniently forget about the physical structure of the manipulator mechanism and concentrate only on the location of the end-effector.

Implementing such a system requires measuring the amount by which you move the control knobs to determine the desired Cartesian position and then converting that desired motion into a feasible sequence of joint motions. To do that, we need to compute the mappings between joint and Cartesian spaces.

2.4 Forward and inverse kinematics

Given that we know the length of each link, and that we may measure the angle by which each joint has rotated, then we can compute the location of the end-effector in Cartesian coordinates. This mapping between joint space and Cartesian space is called *forward kinematics.* For robots such as this one, in which the links are arranged in a serial chain, the forward kinematics computation is relatively simple. Given the length of the first link and position of the first joint, we may compute the location of the end of the first link; then, given that position, we can repeat the procedure for each subsequent link, at each step determining the position of the link one step further from the starting point, until finally the end position of the

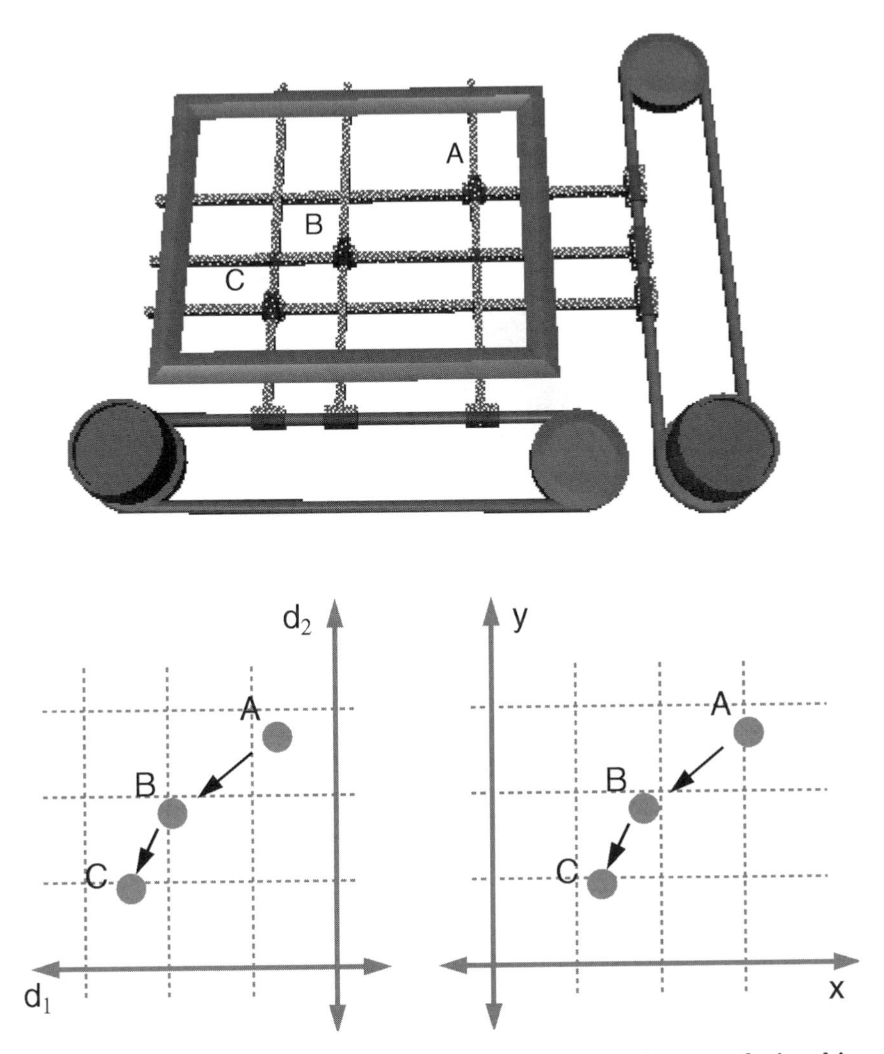

FIGURE 2.4: The Etch-a-Sketch™ has a convenient linear relationship between joint space (d_1, d_2) and Cartesian space (x, y).

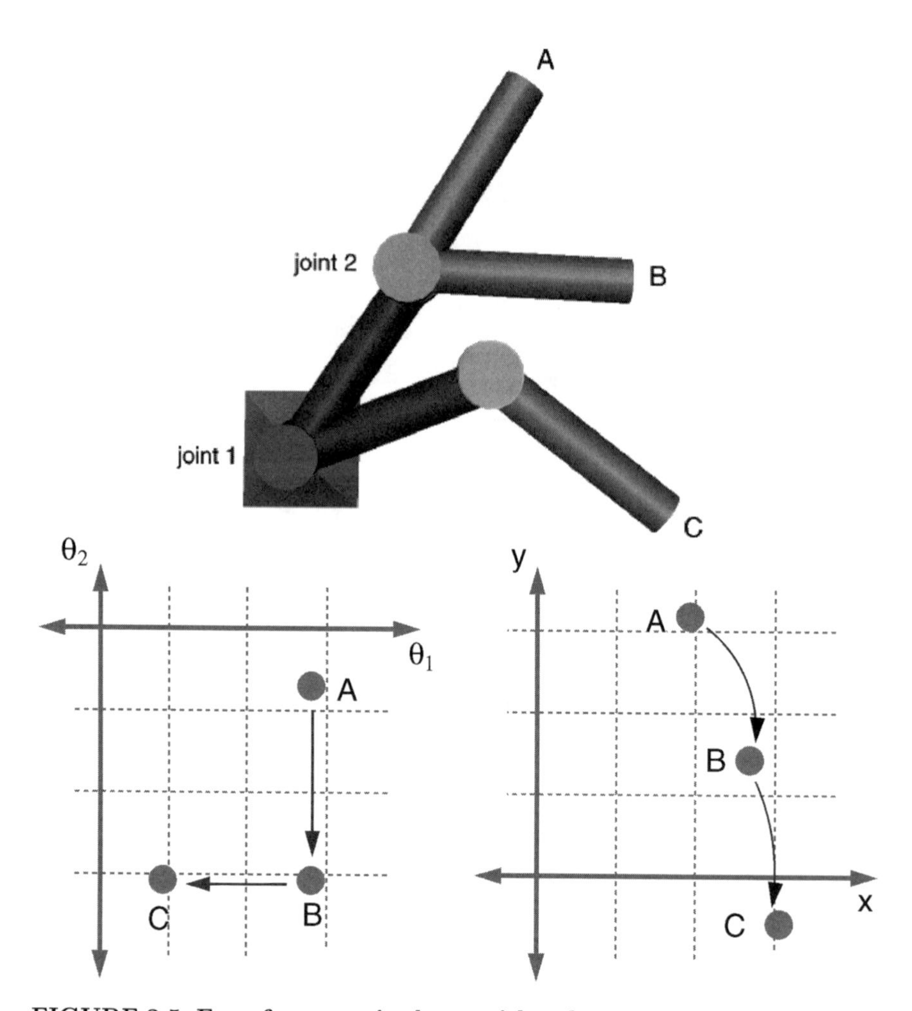

FIGURE 2.5: Even for a manipulator with only two revolute joints the mappings between joint and Cartesian space are not simple. Shown here are both joint and Cartesian space representations of the same simple motion.

robot is determined. These computations may be performed in a simple and elegant manner using a matrix formulation known as *homogeneous transformation mathematics* [59].

For serial manipulators, the more difficult task comes in performing the *inverse kinematics* calculation where we must derive a mapping between Cartesian space and joint space. While it is possible to solve such problems numerically, it is usually considerably more efficient and satisfying to attack the problem with a combination of trigonometry and intuition to determine an exact analytical solution. Solving the inverse kinematics for a new type of robot used to be considered such a sufficiently significant achievement that it consumed whole research papers. However, now the basic principles are understood [60], and this knowledge, combined with the use of modern symbolic math packages [96], and the fact that most modern manipulators are specifically designed to have relatively simple inverse kinematics solutions, makes the task relatively simple.

2.5 Redundancy

Now we have two different means of representing the robot's position symbolically. There is the point in joint space corresponding to the current manipulator position, and there is the point in Cartesian space corresponding to the current end-effector location. For both the robot on your roof and the Etch-a-Sketch™ toy there were simple one-to-one mappings between these different representations. Unfortunately, for most real manipulators the situation is not so simple.

In general, there is often more than one set of joint positions for each desired Cartesian end-effector position. Even in the case of our simple two-link manipulator, there are often two possible solutions, and the system exhibits a degree of *redundancy* (see Figure 2.6).

For the two-link serial manipulator, there are at most two possible inverse kinematics solutions. However, it is possible to have an infinite number. For example, if an extra link were to be added to the manipulator, then for many end-effector positions we could choose any value for the first joint and yet still place the end-effector at the desired location by choosing appropriate values for the remaining two joints. In such cases, more complex additional constraints are possible. For example, we may choose joint values so as to maximize the force that the manipulator may exert in a particular

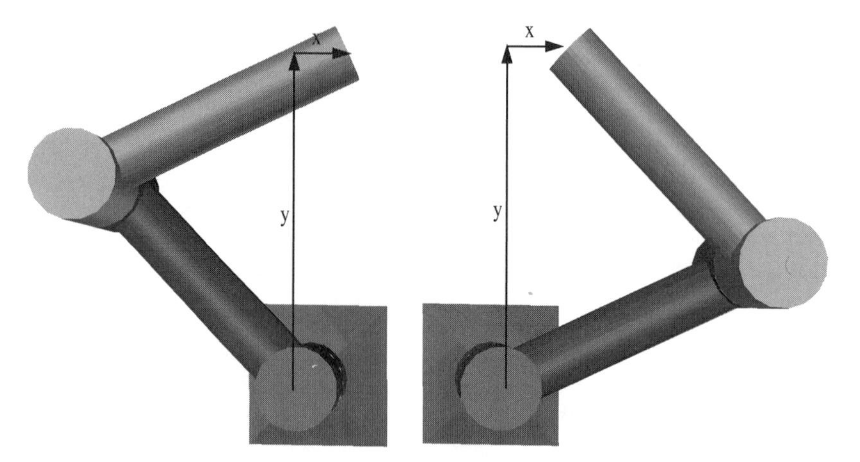

FIGURE 2.6: A two-degree-of-freedom manipulator with revolute joints exhibits redundancy. In this case, there are two different joint positions that yield identical Cartesian end-effector positions. One common means of handling redundancy is to introduce additional constraints. A simple approach is to select from among the possible inverse kinematics solutions by choosing the one that may be reached with the minimum joint space motion.

desired direction. Or we may choose those positions that place each joint as far as possible from any physical joint limit.

2.6 Moving out of the page

All the examples so far in this chapter have been *planar* manipulators whose end-effectors could only move along at most two Cartesian axes. The more interesting, and practical, situation is where we desire manipulators that can simultaneously control both position and orientation; and which can move, not just in a plane, but in a three-dimensional volume. In general, we desire an ability to control the roll, pitch, and yaw of objects being manipulated, while simultaneously controlling their position within a three dimensional Cartesian workspace. Thus, most general-purpose robot manipulators have at least six degrees of freedom. The most famous example of such manipulators in the the robotics literature is the Puma 560 factory robot (see Figure 2.7). This uses six revolute joints; three to control the position of the end-effector, and a further three to control its orientation.

If we examine just the last three joints (comprising the Puma's wrist), then it should be clear that these provide control over all three possible Cartesian axes of rotation (see Figure 2.8). However, a special case occurs if the axes of rotation for the first and last wrist joints are aligned. In this configuration, rotating the first joint has the same effect on the end-effector orientation as rotating the last. Thus, while the wrist usually has three independent rotational axes, in this special case it has only two. The result is that a rotation perpendicular to the middle joint axis is not possible while the wrist is in this configuration (see Figure 2.9). Such situations are known as *kinematic singularities.* Even the two-link manipulator example exhibits such a singularity. If the links are both aligned, then motion of the end-effector directly toward the manipulator base is no longer possible. More generally, if we were to write an equation linking motion in Cartesian space with the corresponding motion in joint space, then singularities correspond to cases where that equation is ill-defined.

Just as for the two-link manipulator, the Puma robot exhibits some redundancy. However, unlike the two-link manipulator, some can be difficult to visualize. For example, Figure 2.10 shows an example of the wrist redundancy. In general, the Puma has as many as eight different joint-space points corresponding to a single point in the Cartesian workspace.

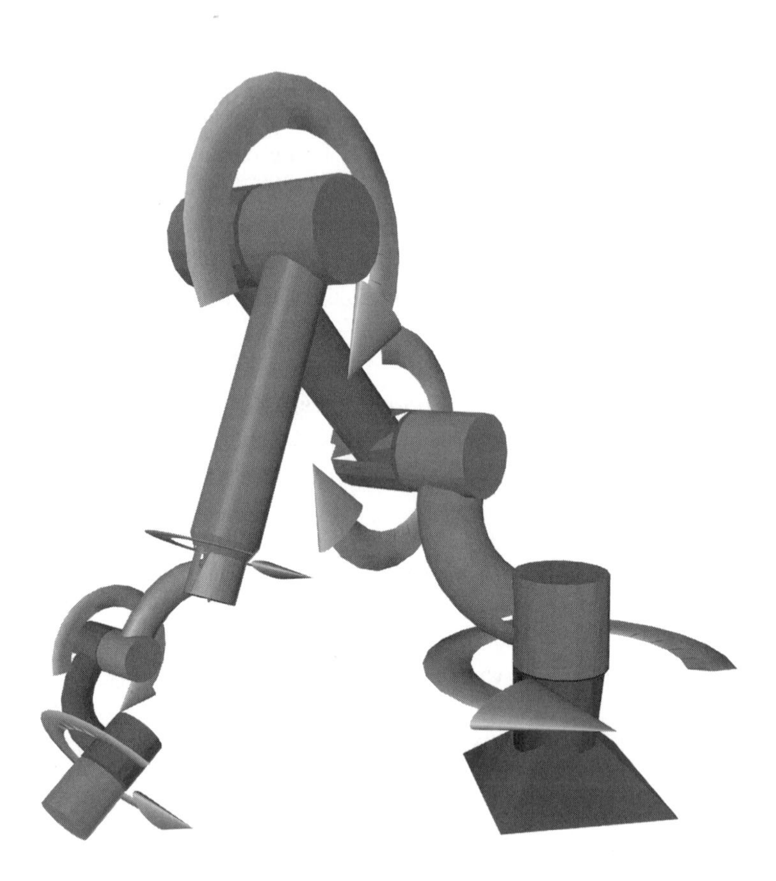

FIGURE 2.7: The kinematic structure of the Puma 560 robot. The first three joints control position, while the last three control end-effector orientation.

FIGURE 2.8: Stylized view of a three-DOF robot wrist as used in the Puma560 factory robot. This style of wrist design is particularly common among commercial manipulators. Inverse kinematics is simplified by having the axes of rotation for all three joints intersect at a single point.

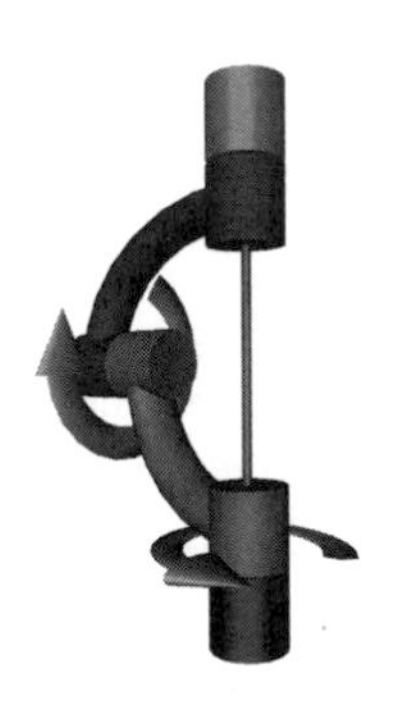

FIGURE 2.9: Stylized view of a three-DOF robot wrist showing the singular configuration when the first and last joints are aligned. In this configuration only two independent rotations are possible.

FIGURE 2.10: Animation of motion between two equivalent Cartesian positions showing redundancy. Two different joint configurations correspond to identical Cartesian orientations and motion between these two alternative configurations passes through a singularity.

2.7 Input devices

Consider the different levels required in order to create a working system. At a minimum, the system must provide the operator with the means to control the remote robot. In general, input devices fall into three categories: allowing binary, rate, and proportional control of joints at the remote site.

2.7.1 Binary controls

Binary controls are simple on/off buttons or levers. Each lever controls a particular degree of freedom. Examples are a car windshield wiper lever, the button-box used to control factory robots, or the on/off switch on a blender. In each case, the operator has control over whether or not a controlled degree of freedom moves, but he or she can't control the speed of motion. And a particular desired position may be obtained only by activating and deactivating the controlled joint at just the right times.

2.7.2 Rate controls

Rate controls are levers, dials, or other input devices that offer a continuous analogue input and where that input controls the speed of the controlled degree of freedom. In these cases, the operator now has control over the rate of motion of the controlled degree of freedom, but control over position still requires activating and deactivating the DOF at just the right times—though it is a little easier since he or she can slow down when nearing a desired location.

2.7.3 Proportional controls

Proportional controls are again analogue input devices, but now the position of the input device directly controls the position of the output device. Examples of this type of control are a car steering wheel or the trackball on the portable computer used to write this chapter. In this case, the most natural of the interfaces, the operator has control over both the position and the speed of the controlled DOF. He or she can not only move it to a desired position, but can do so while controlling velocity.

2.7.4 Computer control

The above represent the three traditional types of interface devices. However, now that a computer often connects the input and output signals, more complex and abstract controls are possible. For example, modern computer mice use a combination of rate and proportional control. At slow speeds, the distance the pointer moves on-screen is proportional to the distance the mouse moves. But, if the operator moves the mouse at higher rates, then the pointer moves increasingly greater distances. Thus, a small slow motion no longer produces the same final location as a small fast motion.

Consider the case of a factory robot. The simplest interface is joint-space control. In this case, each button causes motion of one joint in one direction. For most factory robots there are six independent joints and each can move forward and back, resulting in a total of twelve buttons. By appropriately activating these controls the operator may position the robot at almost any desired position and with almost any desired orientation.

A more sophisticated implementation introduces an additional level of abstraction between human and machine. The idea is to provide the operator with controls for natural motions "move up" "move left" rather than the previous "rotate joint 2 clockwise" style of interface. This is Cartesian control. In some sense, it is more powerful, since now the operator may command coordinated motions of several joints on the remote manipulator, merely by pressing a single button on the control panel. For example, with the Puma robot, when the operator commands the manipulator to move "up" joints 2 and 3 must both rotate (but in opposite directions and with different speeds) and, if the end of the robot (the end-effector) is to maintain a constant orientation, then joints 4, 5, and 6 must also move (again in different directions and with different speeds). This style of interface (see Figure 2.11) permits the operator to think and command the robot in a natural manner while the robot controller assumes some of the complexity of task performance.

An alternative to buttons and levers is to use a second robot as the input device. In this case, the operator holds and moves one robot manipulator. The motions of that arm are duplicated by a second robot arm at a remote location. This is how traditional teleoperation systems work, and they are the subject of the next chapter.

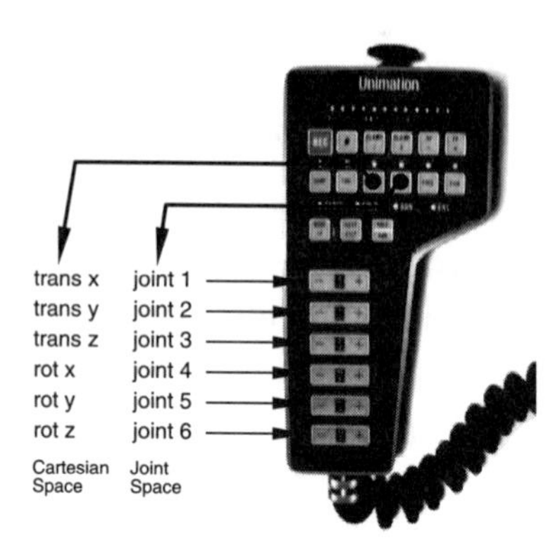

FIGURE 2.11: The teach pendant for a six-degree-of-freedom Puma robot provides a simple example of an input device for both joint and Cartesian control. Cartesian space is more natural; however, the inevitable presence of singularities in the slave robot workspace means that Cartesian motions are not, by themselves, sufficient. It is therefore appropriate to permit either joint or Cartesian motions and allow the operator to choose that mode which is most appropriate at each instant.

2.8 Summary

Regardless of the type of manipulator, all robots have at least two common representations: joint space and Cartesian space. Commanding a manipulator in joint space is simple to implement and powerful to use, but it can be difficult for operators to command straight-line end-effector trajectories. Cartesian control is more natural, but is complicated by the need to compute forward and inverse kinematics, as well as by the inevitable presence of singularities and redundant configurations.

Input devices allow an operator to command a remote robot. They may provide binary, rate, or proportional control. They may be as simple as buttons and levers, or as complex as another whole robot manipulator.

3

Historical Perspective

A number of authors have presented surveys of historical developments in teleoperation [24, 40, 46, 76, 90, 91]. This chapter draws heavily on this material.

Teleoperation allows an operator at one location to perform a task at some other, perhaps quite distant, location. In the case of nuclear processing, just a few centimeters of lead glass separate the operator from the task; while in the case of subsea teleoperation, several kilometers of water may divide the two. However, regardless of the application, all teleoperation systems have the following specific features (see Figure 3.1):

- An operator interface, incorporating a master input device that the operator uses to command the system.
- A slave output device that performs the operator's commanded actions at the remote site.
- A communication scheme between sites.

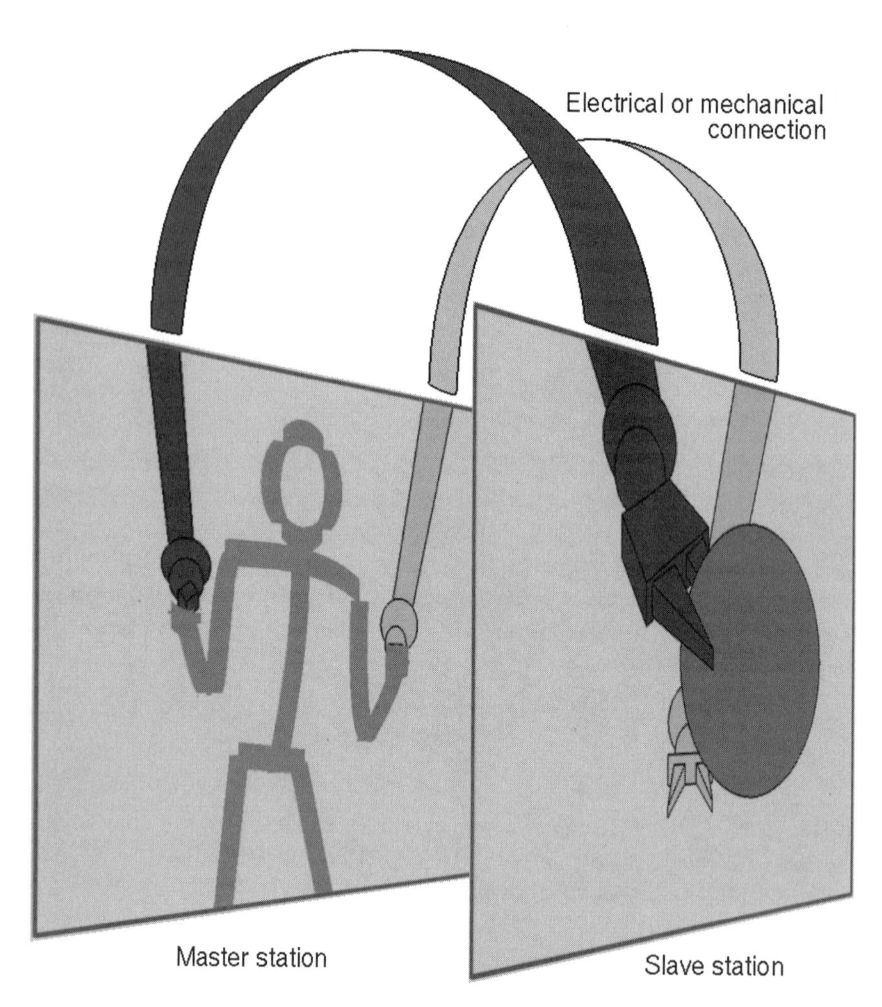

FIGURE 3.1: Teleoperation is the performance of remote work. A human operator at the master station controls manipulators to perform work at a remote slave site. The separation between sites may be as simple as a glass window, or as complex as a satellite link. The connection between sites may be a mechanical linkage or an abstract electrical signal.

We'll begin by looking at traditional teleoperation systems, choosing examples from the movies, the nuclear power industry, and subsea scientific research. Then we'll look at different operator aids that focus on improving efficiency. Finally, we'll look at systems that can cope with a communications delay between operator and remote site. The simplest of these display predicted motions for the operator; more complex systems imbue the slave site with an ability to react without needing to wait for the operator.

3.1 Traditional bilateral teleoperation systems

The first teleoperation systems were developed by Goertz at the Argonne National Laboratory for use in handling radioactive materials. The operator could observe the slave manipulator directly through a thick lead glass window and cause it to move by holding and moving the end of the master manipulator. The designs were complicated by the need for easy removal and repair, employing elegant mechanisms that enabled the entire slave arm to slide through a small round aperture in the radiation barrier.

In these earliest implementations, the master manipulator and remote slave arm were exact scaled copies of each other. They were *kinematically identical.* This guarantees that any singularities or joint limits in the slave manipulator will be in exactly the same place on the master arm. Thus, the operator need only think of the slave manipulator. He or she can imagine they are standing at the remote site, directly guiding the end of the remote arm.

The two robot arms were also mechanically linked. This provided a natural bilateral system in which any forces applied to one robot arm will be felt on the other. This has the desirable effect that when the operator pushes harder on the master arm, the slave arm pushes harder on the environment. When the environment pushes back, the operator feels that impact on his or her arm.

The negative side effect of realistic force feedback is that the operator actually has to do physical work in order to perform work at the remote site. This could be mitigated by adjustable mechanical counterbalancing so that the operator didn't have to constantly support the weight of objects being held at the remote site.

In later systems, the use of electrically powered manipulators removed the need for a direct mechanical connection. This simplified the design with the not insignificant advantage of removing the need for holes in the radiation barrier.

Readers interested in viewing an example of an electrically actuated teleoperation system are encouraged to rent a copy of the movie *The Andromeda Strain* (Universal Pictures, 1970). While this is a work of fiction, the portrayal of remote control robotics is surprisingly accurate.

3.1.1 The MASCOT teleoperation system

A modern commercial example of a bilateral teleoperation system for use in the nuclear industry is the MAnipolatore Servo COntrollato Transistorizzato (MASCOT) system developed by Elsag Bailey (see Figure 3.2) [14]. This features dual six-DOF, kinematically identical, master and slave arms with full bilateral control. Each arm can move up to 20 kg with an accuracy of 0.5 mm. Communication between master and slave sites is via optical fibre cable. The system has the ability to compensate for the weight of grasped objects (so that those constant forces need not be maintained by the operator). It supports reindexing (so the workspace of the slave manipulator may be larger than that of the corresponding master arm), and has a "teach and repeat" function (so sequences of operations may be stored and later replayed).

3.2 Giving up force feedback

In the early mechanical systems, a natural bilateral system was a desirable side effect of the direct connection between sites. With the move to electrical connections, force feedback to the master arm became an optional and costly addition. The powered manipulator at the master site could now be replaced by a passive instrumented linkage. While this is known to be less efficient for at least some tasks [51], it is also significantly less expensive.

The use of an electrical connection also made it possible to do away with the master arm entirely, replacing it by a *button-box interface* in which there was a direct mapping between each input control and each joint on the remote robot. This is the simplest and least expensive implementation. The tradeoff here is that it lacks the naturalness of older interfaces, forcing operators to mentally plan motions and then convert those plans into sequences of button/lever activations.

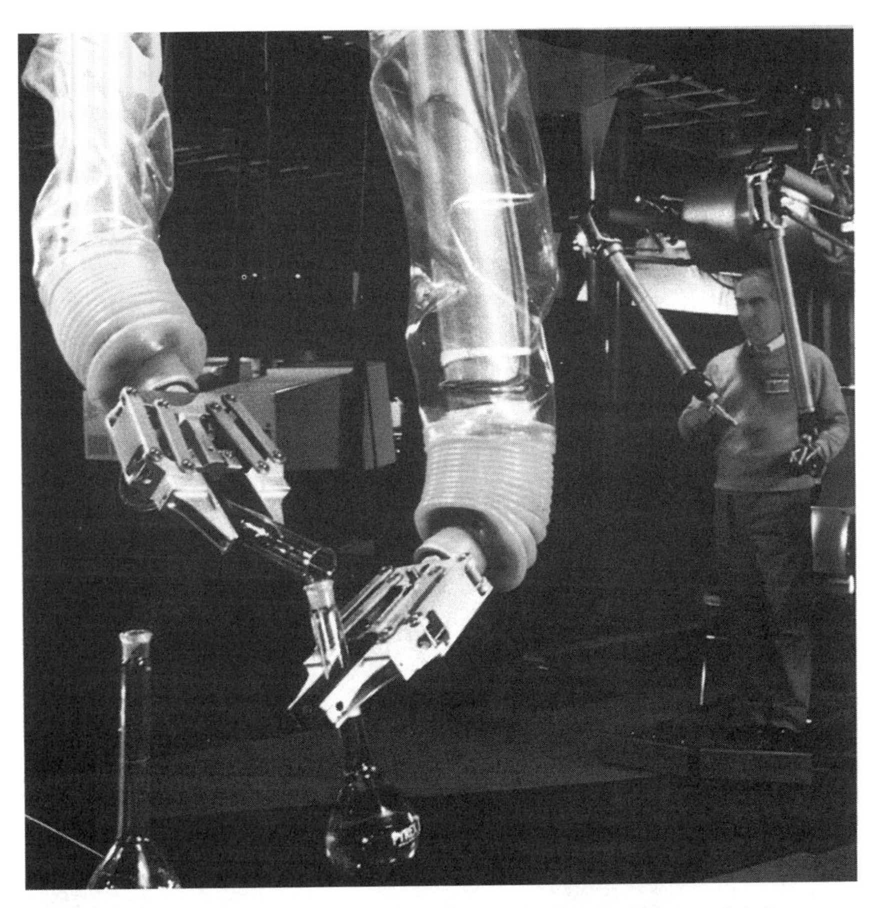

FIGURE 3.2: The MASCOT system—a dual-arm bilateral teleoperation implementation developed for the nuclear industry. Photo copyright Elsag Bailey, reproduced with permission.

An example of a modern system that utilizes both an instrumented linkage and a button-box interface is the teleoperation equipment on the Woods Hole Oceanographic Institution's ALVIN manned submersible.

3.2.1 The ALVIN teleoperation system

Here the operator is protected from the pressures of the deep sea by a titanium sphere. He or she observes the remote world through thick portholes and interacts with it using two robot manipulators mounted on the front of the vehicle (see Figure 3.3). Both manipulators are connected to ALVIN using explosive bolts to enable the submarine and its human occupants to escape should one of the arms become entangled.

The interface for ALVIN's right arm is nothing more than four electrical switches. Each controls a single joint. Move a switch up, and its joint rotates clockwise; move it down, and the joint rotates the other way. Inside the cramped and battery-powered submersible this interface has several

FIGURE 3.3: The Woods Hole Oceanographic Institution's ALVIN manned submersible enables operators within to interact with the inhospitable deep sea environment through two forward-mounted robot manipulators.

advantages. It is very reliable, it takes up little space, and it consumes almost no power. Performing simple tasks is difficult, requiring a skilled operator. Performing delicate tasks is even harder, but for those there is another manipulator.

ALVIN's left arm, intended for more dextrous tasks, is controlled by a passive instrumented linkage. This linkage is essentially just a series of knobs, one for each joint, except that it permits the knobs to be oriented so that they are in approximately the same relative positions as the corresponding joints on the remote arm. By correctly aligning the linkage, operators can imagine that they are holding the remote arm in their hands. In the case of the switches for the right arm, the operator only had control over which direction a joint moved. The knobs for the left arm directly command position; thus, here the operator has control over both position and speed of motion. Skilled operators can move the arm with such dexterity that it almost appears to be alive.

3.3 Other teleoperation systems

So far, every output device we've looked at has been some form of robot manipulator. But this need not be the case. The output device may be any device that performs physical work at a remote location in response to operator input. Turning on a light switch is not teleoperation, nor is accessing a Web page. But operating a toy radio-controlled car is. Other examples are as everyday as building site cranes, or as unique as the JASON remotely operated underwater vehicle (see Figure 3.4).

3.4 Operator aids

As teleoperation systems have moved from a mechanical connection to a reliance on digital technology, the opportunities for assisting the operator have increased. Possibilities for operator aids include supporting reindexing, maintaining a natural interface, providing visual aids, and generating synthetic force clues.

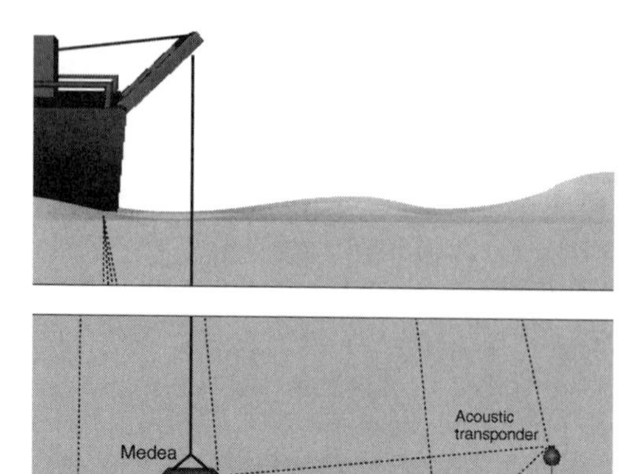

FIGURE 3.4: The Woods Hole Oceanographic Institution's JASON remotely operated vehicle allows operators aboard a surface ship to perform remote work on the sea floor. Communication between humans and remote robot is via several kilometers of heavy armored cable and a short neutrally-buoyant tether. Navigational data is obtained by combining information from depth sensors with the time it takes sound to travel between acoustic transponders and the vehicles. While Medea is essentially just a weight at the end of the cable, JASON is a sophisticated subsea platform. It is equipped with thrusters to control position, sensors to collect scientific data, and a manipulator to gather samples and service instruments on the sea floor.

3.4.1 Reindexing

Imagine an implementation where the master arm has a smaller range of motion than the slave arm. This has the advantage that the master arm can be smaller and cheaper, with the disadvantage that large continuous motions at the slave site can no longer be performed. The trick to making such systems workable is to allow reindexing.

Reindexing is where the connection between master and slave devices is temporarily removed, the master arm is repositioned, and then the connection is reestablished. This allows a master arm with a relatively small workspace to move a slave arm through a large motion by using a series of discrete motions (see Figure 3.5).

The introduction of reindexing is not without cost. Now that the arms may move independently, there's no longer any one-to-one correspondence between singularities and joint limits at each manipulator. The natural correspondence between arms has been lost. For those axes of motion in which reindexing is supported, the operator must now be concerned with limitations of both the master arm and the slave manipulator.

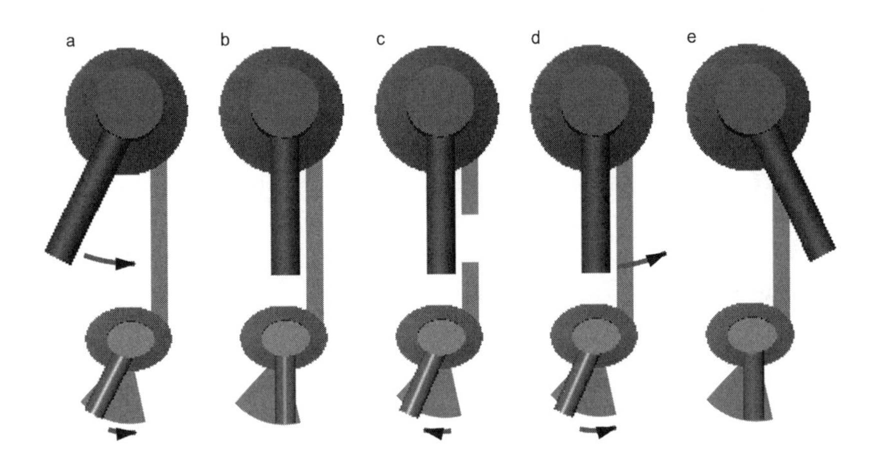

FIGURE 3.5: A master arm with a small range of motion may be used to control a slave arm with a much greater range of motion by using reindexing. The slave arm at top follows the motions of the master arm below (a). When the master approaches a joint limit (b), the connection between sites is broken and the master arm is repositioned (c). This is the reindexing operation. Then the connection is reestablished (d) and motion continues (e).

3.4.2 Maintaining a natural interface

In the old, mechanically linked systems, there was a fixed correspondence between the master and slave devices and the operator could only view the slave from one direction. By approximately aligning the two arms along this viewing direction, it was possible to ensure a natural mapping between operator and slave coordinates. If the operator moved the arm to the left, then the slave would also move toward the operator's left. If he or she moved up, the slave would also move up. With the move to video feedback, this natural correspondence was often lost. It was simpler and, in the case of kinematically similar master and slave manipulators, necessary, to maintain a fixed mapping despite significant changes in viewing direction. This is undesirable. To see why, try this simple experiment. Take the mouse and mouse pad from your computer and rotate them by ninety degrees in the plane of the table top. Now try to move the mouse pointer to a particular spot on the screen. It's much more difficult than one might imagine. Interestingly, some rotations are more difficult than others. For example, the natural position of the mouse pad is rotated ninety degrees from the screen—a forward/back motion of the mouse causes up/down motion of the pointer. Yet most people appear to have no difficulty in performing this mapping.

Experiments on robot control using teach pendants indicate that changing the orientation of the robot with respect to the operator can adversely affect performance [31], and Rasmussen has shown that performance at teleoperative tasks degrades when the operator's frame of reference differs significantly from that of the slave robot [63]. Brooks has termed this a "Stimulus-Response Mismatch" [12].

The solution is to remap the operator's input based on their viewing direction so that natural motions are again possible. This has been known in the area of computer graphics for some time [10]. A simple example of such an implementation is the space shuttle arm. It uses several alternative (but fixed) mappings so that Cartesian controller motions are performed in a natural reference frame for each viewing porthole [91]. A more complex example is IBM's telerobotic camera manipulation system [88]. It allows surgeons to specify the motion of an endoscopic camera by using natural viewer coordinates ("move left", "move up"). Translating these motion requests into robot commands is nontrivial due to the motion constraints imposed by the patient's body [29].

3.4.3 Automatic camera control

The use of video cameras at the remote site makes it possible to provide views of the remote task site from several different directions. Remote site cameras may be mounted on pan/tilt or other mechanical linkages and may be automatically positioned based on operator head motion. The earliest example of this was the Goertz implementation [90], while a sophisticated modern example is the stereo vision telepresence system developed by Tachi et al. [58]. Cameras may also be controlled to follow the motion of the slave robots [12, 14] or, more recently, to pan/tilt/zoom based on the particular task action being performed at the slave site [92].

3.4.4 Computer-generated imagery

The availability of high-resolution, real-time, computer-generated displays has made it possible to supplement or replace traditional raw video imagery. Kim and Stark [43] overlayed visual enhancements on real video imagery showing a stick figure model of the robot hand, and found a significant decrease in completion times for a pick-and-place task. A more recent example included full CAD models of the remote site shown in wireframe form over the real imagery [58].

By overlaying computer-generated imagery, it is also possible to display predicted arm motions [8, 34, 44, 78]. (This has implications for teleoperation in the presence of delay—see later.) It might appear that a computer-generated interface would offer inferior performance to direct vision. However there is some evidence to suggest the reverse. Rovetta et al. compared different operator interfaces by measuring electromyographic (EMG) response [67]. They concluded:

> "...in the direct view condition the neuromuscular response was a little slower than in the virtual environment. With a direct visual feedback the subject has to deal with a larger amount of information, not all of it relevant for performance of the task. Thus, although virtual reality offers an impoverished view of the world, it may lead the operator to better focus on the relevant information."

Another advantage of computer-generated imagery is that it makes it possible to break the temporal dependency between master and slave sites.

Operators may work either faster or slower than the real slave robot. Conway et al. have introduced the notions of the *time clutch* (where the operator may work ahead of the real robot while making use of a predictive display), the *position clutch* (where the operator may test motions of the predicted slave without any commands being sent to the real slave), and the *time brake* (where motions performed in advance of the slave robot are successively discarded) [15, 16].

3.4.5 Kinesthetic aids

The first kinesthetic aids for teleoperation were appropriately scaled physical models of the real remote environment. The operator could command the slave by simply positioning the master arm within the model. Since any discrepancy between the model and real world could potentially be disastrous, great care was taken in ensuring consistency [64].

> "...the visual checks of the operator are complemented by several feedback sentinels: (i) tracking error for each servo in the slave... (ii) position offset during tool coupling and release, and (iii) binary switches to verify motion sequence completion. These built-in checks give the system operators the detailed measurement data they need to detect and correct subtle shifts in the world as they occur. "

More recently, improvements in computer vision have made it possible to consider using the operator's hand to directly command a remote robot. The operator performs the task in the real world, using real objects, and the system infers a robot program by observing his or her actions [39]. This is an innovative approach. However, it is complicated by the need to implement a machine vision system that can recognize operator motions and limited by the need to operate in the real world with little opportunity for automated aids to assist the operator.

The use of synthetic, computer-generated force feedback was suggested by Sutherland as early as 1965 [84]:

> "The force required to move a joystick could be computer controlled. . . With such a display a computer model of particles in an electric field could combine manual control of the position of

> a moving charge, replete with the sensation of the forces on the charge, with the visual presentation of the charge's position. "

A system very much like this description has since been implemented [57], and it has been reported that "haptic-augmented interactive systems seem to give about a two-fold performance improvement over purely graphical interactive systems" [11]. This is perhaps not surprising since similar improvements have been found for conventional teleoperation with and without force feedback [51].

Computer-generated force feedback has also been used to simulate collision forces in both predictive teleoperation and pure virtual reality applications [3, 55, 42, 85]. These all adopt similar approaches—modeling the surface as a spring-damper and generating forces whose magnitude increases with increasing object penetration. The aim is to increase the operator's "sense of presence" by providing a sense of touch as well as one of vision.

In generating collision forces, the intention is usually to provide realistic force feedback; however, several researchers have also considered unrealistic force clues. These have been used to repel the end-effector from obstacles in the environment [13], to constrain end-effector motion in one or more degrees or freedom [89], to confine the boundaries of the end-effector trajectory [66] and, most recently, to create synthetic input/output devices [36].

3.5 Increasingly inter-site distances

In the earliest teleoperation systems, the distance between operator and remote site was constrained by the need for the operator to directly view the remote environment. We'll term these systems *short-range teleoperation*. In this case, there is no constraint on the flow of information between sites, and there is no communications delay. The operator may readily imagine that he or she is actually in the remote environment, and there is nothing to break that illusion.

The obvious next step was to combine an electrical teleoperation system with a means to permit viewing of the remote site from a distance. The addition of television cameras and monitors meant that the separation between operator and remote environment could be greatly increased. These systems enable *medium-range teleoperation*. In such systems, the connection between sites is entirely electrical and there is no perceptible communications delay. The operator may see what is happening via a camera and television monitor, he or she may hear what is happening via a microphone and speaker, and he or she may feel what is happening via a force-reflecting master manipulator.

3.5.1 Coping with delays

As the distance between sites increases, the communications delays increase to the point where conventional force-reflecting teleoperation systems fail. This includes all those systems where the master input device and slave robot are connected in a closed control loop. Such a closed system is advantageous without delays since it allows forces felt by the slave arm to be reflected back to the operator. However, it fails in a delayed environment because of inherent instability [77].

In some cases, it is possible to operate in an open-loop fashion where forces are not fed back to the master arm. Alternatives to direct force feedback are indirect feedback, where sensed forces are fed back either to the operator's other arm [91], or as sound [5, 91], or as a graphical display [7], or no force feedback at all with a visual display providing the primary information channel.

These systems tolerate delays by using the human operator as an intelligent link in the control system. Any error in the position of the remote manipulator is visible in the returned imagery, and the operator moves the master arm to compensate.

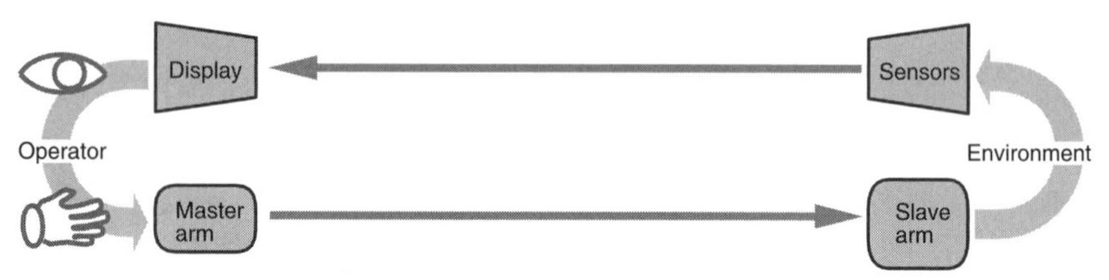

While open-loop systems are certainly not optimal, they are simple and they work. Being tolerant of large communication delays they enable *long-range teleoperation*. Human operators can perform remarkable feats with even the most primitive of interfaces.

However, as delays become large, the efficiency with which operators can work drops dramatically. It has been shown that, in such situations, operators will tend to adopt "move and wait" strategies [25] in order to minimize the risk of incorrect slave motion. In such cases, the operator makes a small motion, waits to see if it works as expected, and then makes another small motion, waits again, and so on. Performing large or complex actions is discouraged, since the operator has no chance to react should anything go wrong. You saw the problem when you smashed that sphere in the introduction. By the time feedback from the remote site arrives, it is too late to correct any error.

3.5.2 Predictive displays

The way to improve performance is to provide the operator with immediate feedback. Because of the communications delay, that immediate feedback can't come from the remote site. Instead it must be generated at the operator or master station. The idea is to insulate the operator from the communications delay by providing simulated or predictive feedback.

The simpler examples merely show where the slave arm will move [8, 78]. More complex systems consider kinematic models of the remote environment [44, 49]. In rare cases, where very detailed knowledge exists of objects at the remote site and where there are no external influences, then it is even possible to consider dynamic models of the remote environment [34].

3.5.3 Increasing remote intelligence

The predictive display's main role is to assist the operator in visualizing what will happen at the remote site. But it is not enough. The communications delay means that we can no longer rely on the operator's reflexes to detect and correct problems. Instead we must provide the remote site with sufficient artificial intelligence to enable it to react to problems immediately. The goal is not to make a robot that's as intelligent as a human. Rather, the goal is to take a few well-defined low-level reflexes and encode those into the remote site.

Now, given a predictive display at the operator station and some low-level reflexes at the slave, we have the means to cope with communications delays. These systems allow the operator to adopt a more supervisory role and they fall under a general category known as *Supervisory Control* [76].

But we must also cope with two other factors: a very constrained communications scheme and a very real and unpredictable remote environment. To cope with those we'll need something more, and that is the subject of the following chapters.

3.6 Summary

Teleoperation systems have progressed from mechanical to electrical and finally to modern digital implementations.

Early electromechanical systems were simple but surprisingly effective. Because of the direct connection between master and slave arms, they provided bilateral feedback with natural kinematic correspondence. Later systems have often compromised the naturalness of those early interfaces in order to operate over larger distances or with lower cost.

Techniques for improving performance include supporting reindexing, maintaining a natural interface, providing visual overlays, automatically controlling remote cameras, and generating kinesthetic feedback.

Supporting long-range teleoperation efficiently requires providing predictive displays and giving the remote robot a low level of local autonomy. The display helps insulate the operator from the communications delay, while the remote autonomy allows the slave to react without needing to wait for the operator.

4

Remote Control

When a direct connection exists between operator and remote sites, then bandwidth is virtually unlimited (if you can run one wire, you can easily run another), and there is no perceptible communications delay. Under these conditions there is no cost to sending information between sites. Its worthwhile transmitting every video frame and every bit of force feedback just in case it might be helpful.

When the direct wire is replaced with a more indirect connection, such as the Internet, then the nature of communication changes. Now there is a limited bandwidth and communications are delayed. In addition, since it is a shared resource, there is an advantage in not sending information, even when it may be useful at the other site.

This chapter introduces the reader to the problems of constrained communications. By way of explanation, we'll pick some examples of long-range teleoperation from the realm of the Internet and, in particular, the World Wide Web. Implementations in this domain must cope with delays on the order of a few hundred milliseconds and bandwidths on the order of a few hundred kilobits per second. These systems are relatively new and rather simple in comparison with traditional teleoperation; however, they have the advantage of being accessible to millions of users. As in life, popularity is often more important than functionality.

We'll begin with the transmission of images from a remote site and then move on to consider the control of remote machines and sophisticated robot manipulators.

4.1 Control of remote cameras

The control of remote camera systems via the constrained communications of the World Wide Web is the most popular remote control technology available via the Internet.

Early implementations allowed remote users to watch a coffee pot at the University of Cambridge[1] and a fish tank at Netscape Communications Corporation.[2] One excellent recent example is the VolcanoCam—an Internet camera that allowed users throughout the world to view live images of Mount Ruapehu, an active volcano in the central North Island of New Zealand.[3] Figure 4.1 shows the operator interface. Users request a new image by simply clicking on the old picture.

The VolcanoCam takes a new image every minute and uploads it via a cellular modem to the Web server. What happens after that is constrained by the limited bandwidth of the Internet.

Dealing with bandwidth constraints requires three steps:

- Reducing the information content.
- Encoding it efficiently.
- Sending it infrequently.

The VolcanoCam's designers reduced the information content by choosing a slow frame rate and a small frame size. They encoded each frame efficiently (and further reduced the information content) by using a lossy JPEG image compression algorithm [62]. Finally, they reduced the amount of information transmitted by sending images only on demand.

The use of compression allows us to trade off between the computational burden at each site and the bandwidth requirements between sites.

---

[1] http://www.cl.cam.ac.uk/coffee/coffee.html
[2] http://www.netscape.com/fishcam/fishcam_about.html
[3] http://www.cybercorp.co.nz/ruapehu

"...And then I started to think. I could now go home, open my laptop, dial in over two loosely bound pre-historic copper cables, and get live pictures in my living room of a mountain four hours' drive away."

—*Julian Meadow*

FIGURE 4.1: The VolcanoCam is one of the first, and certainly most unique, examples of transmitting live imagery from a remote site via the Internet using a WWW user interface. The interface is shown on the right. Shown at left are the images that users saw on July 6th, 1996, while Mount Ruapehu was erupting. Thanks to Julian Meadow and CyberCorp (NZ) Ltd, for allowing us to reprint these images.

The communications delays imposed by Internet communication are just as much a problem as the limited bandwidth; unfortunately, they are much harder to overcome. One technically possible design is to send images continuously to anyone who happened to have the VolcanoCam web page open. In this case, communication is unidirectional and, aside from the delay for the first image, communications delays have no discernible effect on the user's experience.

However, bandwidth on the Internet is currently in high demand and so, while this is technically possible, it's socially inadvisable. Instead the system only sends a new image when the user explicitly requests one.

In the case of the VolcanoCam, all tradeoffs were made at the design stage and hard-coded into the system. In later chapters, we'll look at some mechanisms for dynamically trading off limited resources.

Of course, the viewing of live images from a remote site is not, in itself, remote control—one cannot influence the outcome of a live television broadcast merely by sitting in front of the screen—however, visual imagery is a powerful and very natural means to transfer information on the state of the remote site to the operator, and it will form a part of most remote control implementations.

4.2 Controlling a remote machine

Perhaps the most interesting examples of remote control via the Internet are those that allow users to actively interact with a remote environment. Here the effects of delays become more pronounced and the solutions more demanding.

In this section, we'll look at three different examples of remote control using a World Wide Web interface. The first is a robot archaeologist, the second a toy railroad, and the third a six-degree-of-freedom industrial manipulator.

4.2.1 Robot archaeologist

One of the earliest implementations was the Mercury project at the University of Southern California.[4] This allowed operators to explore a toy

[4] http://www.usc.edu/dept/raiders/

archaeological site by controlling a remote robot. The user could direct the robot through a natural Cartesian interface and command it to blow puffs of air, disturbing the sand to reveal artifacts hidden at the remote site.

This system was online from August 1994 through March 1995. In that time, over 2.5 million hits to the site were registered.

The operator interface for the Mercury project is simple and elegant (see Figure 4.2). Users interact with two pictures. The first shows an image from a camera on the remote manipulator. Here the operator indicates the point in the current image that they would like to be in the center of the camera view after the motion. For larger moves, where the desired end-point may not yet be visible on-screen, there is a second image showing a top-down view of the entire robot workspace. Once again, the operator merely clicks on the point to which they would like the remote manipulator to move. In addition to the clickable maps, the operator may raise or lower the remote arm and may command it to blow a puff of compressed air. The interface works well because most motions are confined to the horizontal plane. The system is reliable because its interaction with the environment is limited to puffs of air.

Once again, constrained bandwidth communication is accommodated by sending small compressed images. The elegant interface allows relatively high-level motion commands to be sent from the operator to the remote site. The limited interaction with the remote environment makes it unlikely that unexpected events will occur—there's only so much damage you can do with a brief puff of air.

4.2.2 Toy railroad

Another example is the Interactive Model Railroad from the University of Ulm in Germany.[5] This allows anyone with a World Wide Web browser to control a model railroad and observe the result (see Figure 4.3).

In this case, the limitations of delayed, low-bandwidth communications are handled by sending relatively high-level commands from the operator to the remote site and by sending small, compressed images back. Unlike a home train set, it's not possible to vary the locomotive's speed or suddenly

[5] http://rr-vs.informatik.uni-ulm.de/rr/

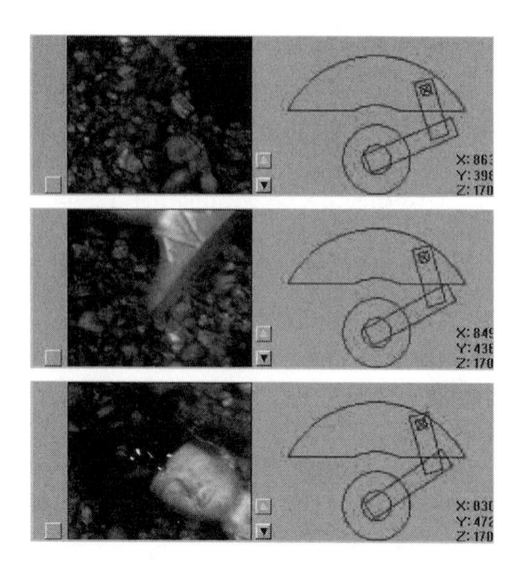

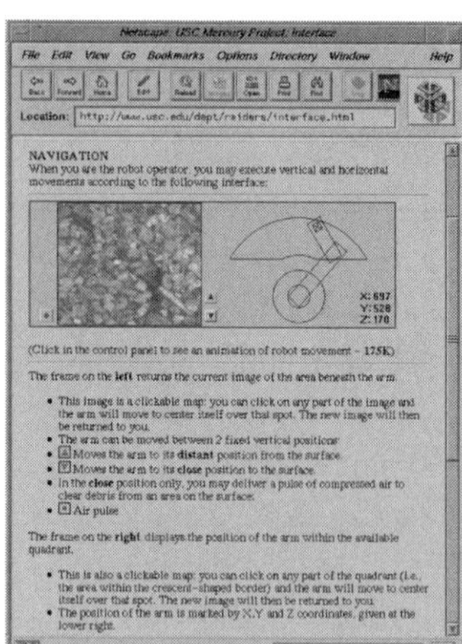

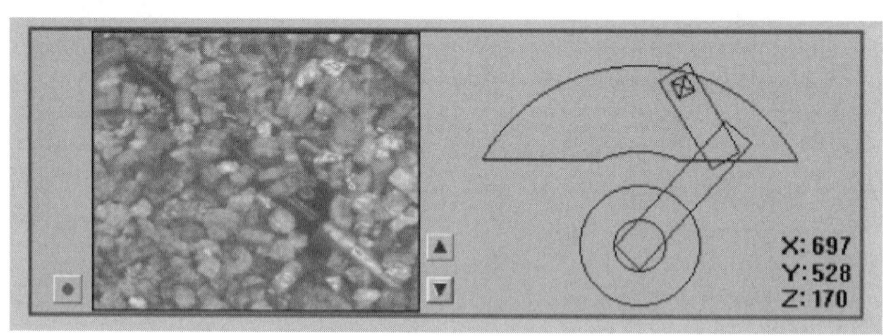

FIGURE 4.2: The Mercury Project at the University of Southern California allowed remote visitors to explore an archaeological site. The three images at top left show an example motion; the full interface is shown at right with an expanded view below. Note the two clickable images that form the primary operator interface. The left image shows a view from the remote arm-mounted camera, while the right image shows a stylized view of the robot's workspace. Thanks to Ken Goldberg for allowing us to reprint these images of his system.

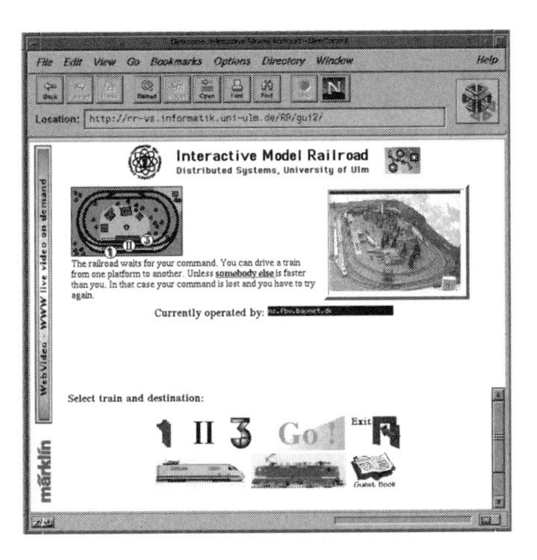

FIGURE 4.3: The Interactive Model Railroad goes one step further than the VolcanoCam—it not only shows users images of a remote site, but also allows them to interact with it—controlling the operation of a model railroad. Thanks to Heiner Wolf and Konrad Froitzheim of the University of Ulm, Germany, for allowing us to reprint these images of their system.

change direction. Instead, one sends commands of the form "make train 2 go to platform 3."

Here the designers have raised the level of communication between the operator and the remote site. Instead of sending dozens of messages a second, the system instead is transferring only one command every few seconds. This idea—of overcoming bandwidth constraints by communicating at a more abstract level—is fundamental to remote control via constrained communications.

Of course, since the trains are real objects, they are unfortunately subject to physical constraints. They may collide or derail and, in such cases, while the operator may be able to see the problem, he or she is unable to do anything about it.

The difficult part of remote operation is not the transferal of commands to the remote site, nor is it the interpretation of those commands as physical actions. Instead it is the development of mechanisms to detect and overcome those problems that will inevitably occur. The need to work despite failures will be revisited many times in this text.

4.2.3 An industrial manipulator

Researchers at the University of Western Australia have made a five-axis industrial robot available for use on the Internet via a World Wide Web interface.[6] Here the remote site is considerably more complicated, the robot has six degrees of freedom (five motions plus the gripper), it is observed by four cameras, and it has a table full of toy building blocks within its workspace.

The operator interface (see Figure 4.4) shows views from each of the remote cameras and accepts commands in the form of Cartesian end-effector coordinates. This is more powerful, but somewhat less elegant, than the previous examples.

Once again, small compressed image fragments are sent from the remote site to the operator. However, in this case the operator has a number of manual controls in order to trade off between the amount of visual information and the rate at which it is updated. He or she may disable cameras, vary the size of each image, and even vary the number of discrete grey levels used to encode each image.

This system provides the operator with manual control of every detail, from the position of the end-effector to the number of greyscales used to transmit each image. This gives him or her considerable power and flexibility. However, that power is not without cost. In some sense, every additional control imposes a tax on the operator.

One solution is to follow the lead of conventional teleoperation by adopting predictive displays and automated camera control. Another is to raise the level of communication between operator and remote sites, providing the remote site with enough information to enable it to make some decisions autonomously without needing to bother the operator. A system that does this is the subject of the next chapter.

4.3 Summary

The limitations imposed on the VolcanoCam by Internet communication—low bandwidth and delayed communications—will occur repeatedly through-

[6] http://telerobot.mech.uwa.edu.au

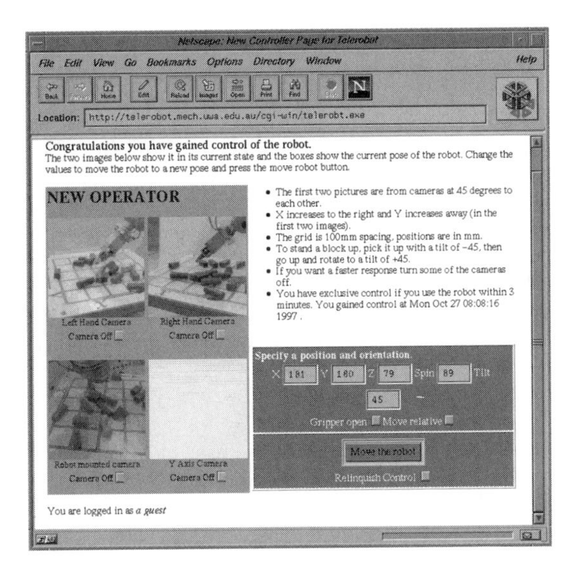

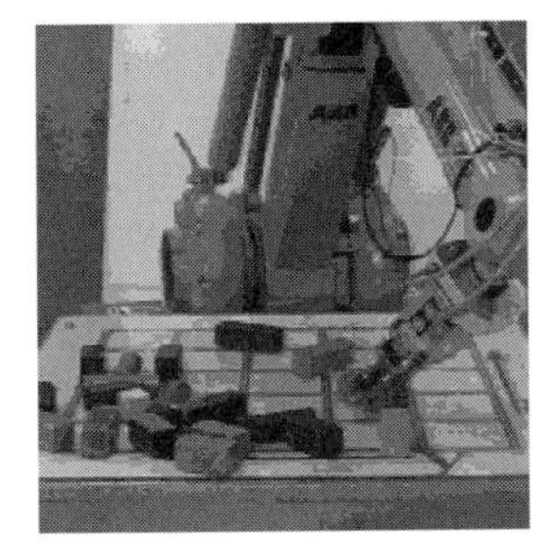

Begin with gripper open, $x = 150$, $y = 150$, $z = 7$, $spin = -45$, and $tilt = 45$.

Close the gripper, move up by setting $z = 100$, then reorient gripper with $x = 200$, $y = 200$, $spin = -89$, and $tilt = -45$.

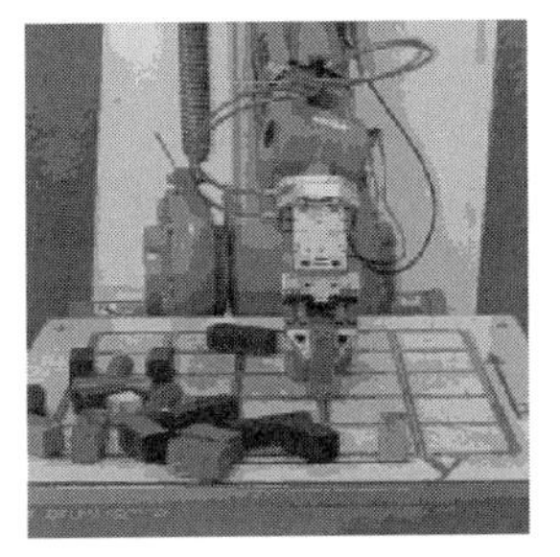

Now lower the block by setting $z = 37$.

FIGURE 4.4: The University of Western Australia has one of the most sophisticated robots on the Internet. Shown here is the user interface for commanding the robot and an example sequence of commands for repositioning a block. For more information see http://telerobot.mech.uwa.edu.au. Thanks to Ken Taylor for allowing us to reprint these images of their system.

out this book. In this case, they are handled by transmitting images only on demand, by choosing a small image size, and by using JPEG compression.

Existing systems for teleoperation via the Internet range from the simple and elegant (the Mercury project), to complex and powerful (the UWA robot). Ideally we'd like a system that provides the simplicity of the former while still providing the power of the latter. The key to achieving this is to build on existing research in conventional teleoperation and to raise the level of communication between sites.

It's certain that teleoperation via the Internet will improve significantly over the next few years. Some advances that can be expected include overlaying visual clues, showing predicted motions, and automating the control of remote cameras.

5

Teleprogramming

In the early chapters of this text, readers were introduced to the basics of robotics, the history of teleoperation, and some of the difficulties of controlling machines via constrained communications. Now it's time to put the pieces together. We desire a system that permits remote control of a robot in a real environment via a constrained communications link. The system that makes this possible is *teleprogramming*, and it is this paradigm on which the rest of this book is based.

This chapter will introduce the principles of teleprogramming by first looking at conventional robot programming and then merging those ideas with teleoperation. To further define the approach, we'll introduce an analogy between a microprocessor cache and a remote robotic system and see how similar tradeoffs occur in both fields.

5.1 Background

The teleprogramming paradigm was developed by Paul, Funda, and Lindsay [28, 30, 47] at the General Robotics and Active Sensory Perception Laboratory of the University of Pennsylvania.

By way of explanation, imagine that your task is to program a factory robot. In this case, the goals are very clear: the robot must reliably perform the same task many thousands or millions of times. Given that goal, it's worthwhile to consider even those problems which may occur very infrequently and explicitly write code to work around each one. This makes sense because the cost of your program can be amortized over many thousands of operations. Programming in this way is an offline process. You complete the program, download it to the robot, and then leave for vacation.

Now, consider how the task changes when you know the program will only be executed a few times. Here the additional burden of every additional piece of error-correcting code becomes very significant—it is difficult to justify coding for something that may only have one chance in a thousand of occurring. Ideally, you'd like to write code only for those problems that will actually occur, rather than having to write code for every problem that might occur.

We saw in the previous chapter that the way to cope with constrained communications is to raise the level of communication between operator and remote site. We can think of those symbolic inter-site commands as being a language and the task of the operator as being that of a programmer. The only difference between this and the situation in the previous paragraph is that as each command is "written" by the operator, it is immediately transmitted to, and executed by, the remote slave robot. Because the operator is writing the program online, we don't need to predict in advance everything that might go wrong. Instead we can wait to see what actually fails and then rely on the operator to code around each error.

This idea, of coping with communications constraints and uncertain environments by using online robot programming, is fundamental to teleprogramming. Rather than have the operator type a robot program, we instead have him or her perform the task in a virtual remote environment and infer a robot program by watching his or her actions. Thus, so long as everything goes well, the operator may imagine that he or she is controlling a conventional teleoperation system with no time delay.

When something does go wrong, an error message is sent back to the operator station, the virtual reality model is reset to correspond to the current slave state, and the operator must continue programming from that new starting point.

5.2 Operation

Teleprogramming works by decoupling the master and slave sites, using local high-bandwidth feedback at each site, and sending low-bandwidth symbolic robot program instructions between sites.

The communications problem is largely overcome by distributing knowledge more evenly throughout the system. Rather than keeping all knowledge of the remote site at the remote site, we instead create a partial copy of that information for use by the master station. Similarly, rather than keeping all task knowledge at the operator station, we maintain a partial copy at the slave station. The result is that the master station can provide immediate feedback for the operator without needing to wait for the slave, and the slave may react immediately to its environment without needing to wait for the operator.

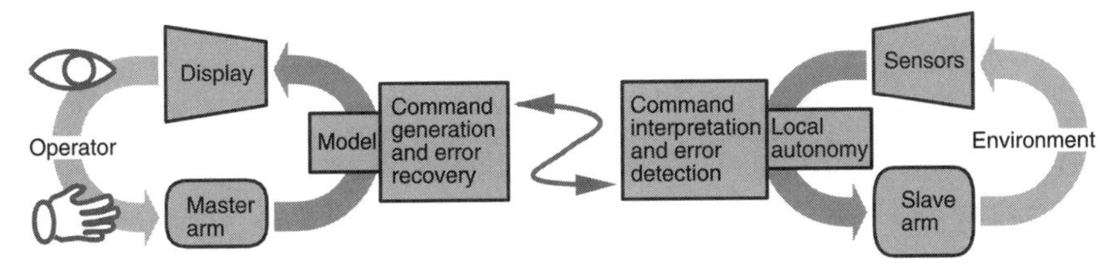

At the operator station, the partial copy of the remote site is used to create a virtual reality representation of the slave site. This simulation provides the operator with immediate sensory feedback. Since his or her interaction is entirely within that simulated world, it is largely insulated from the limitations of the communications channel. In some sense, one can think of this as a cache. Knowledge about the real remote site is fetched, and then subsequent operator enquiries can be handled by fetching from that cache rather than the real remote site. At the slave site, the partial copy of the master station is used to create a level of local autonomy that allows the slave manipulator to react immediately to sensory data.

Transmission between sites is symbolic. The master station observes the operator's interaction within the simulated world and translates that into a symbolic command stream. The slave system interprets and executes those commands generating symbolic reply messages for transmission back to the operator station. By distributing knowledge between the master and slave stations, and by raising communication between sites to the symbolic level, the teleprogramming paradigm permits teleoperative tasks to be performed via a delayed low-bandwidth communications link.

To cope with the unexpected, the system encodes predicted results within the command stream. As the slave executes each command, it compares expected and actual sensory readings, watching for discrepancies. Small differences may be accommodated by mechanical compliance. Large differences may be the result of an unexpected failure (perhaps a part breaking) or an inconsistency in the world model (perhaps an object has moved unexpectedly). When such an unexpected event occurs, the slave pauses and sends a signal to the operator station. It is then up to the operator to diagnose the error and take corrective action, causing new post-error commands to make their way to the remote site.

When all goes well, the operator performs the task within the simulated environment at the master station and, one communications time delay later, the slave performs those same actions. It is not necessary for the operator to wait for the slave to finish one command before sending the next. Since he or she is working within a simulation that reacts immediately, it is possible for the operator to continue several commands ahead of the remote system. Thus, the master and slave systems operate in a pipelined fashion (see Figure 5.1).

When the slave detects an error, it pauses, and sends a signal back to the operator station. That signal takes one communications delay to reach the operator. When it arrives, the operator is interrupted, and the system resets to show him or her the point at which the error occurred. It is then up to the operator to diagnose and correct the problem, generating new commands for the remote robot (see Figure 5.2).

5.3 The fundamental tradeoff

Consider the case where an error occurs at the remote site. A message describing that error must be passed from the slave to the operator station; then the operator must diagnose and attempt recovery from the error and finally, new commands must make their way back to the slave station. Thus, each error has a significant cost.

Sending information also has a cost—it uses bandwidth and computational resources and, perhaps more importantly for some applications, it consumes power. Thus, in practice there is often a tradeoff between the cost of sending additional information and the possible savings if that additional information can avert, or at least mitigate, error effects.

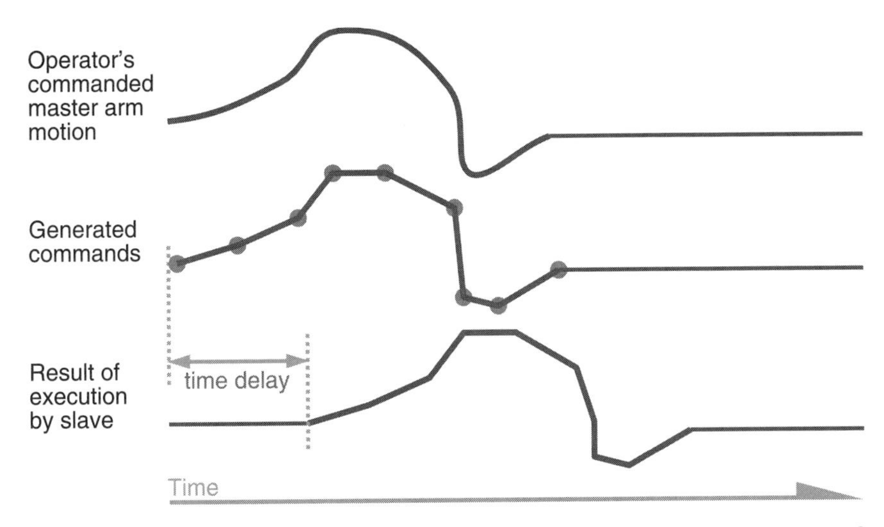

FIGURE 5.1: The operator moves the master arm within the virtual slave environment. That motion is observed and translated into robot program instructions for transmission to the slave robot. Ideally, the operator continues with the task without waiting for the slave robot.

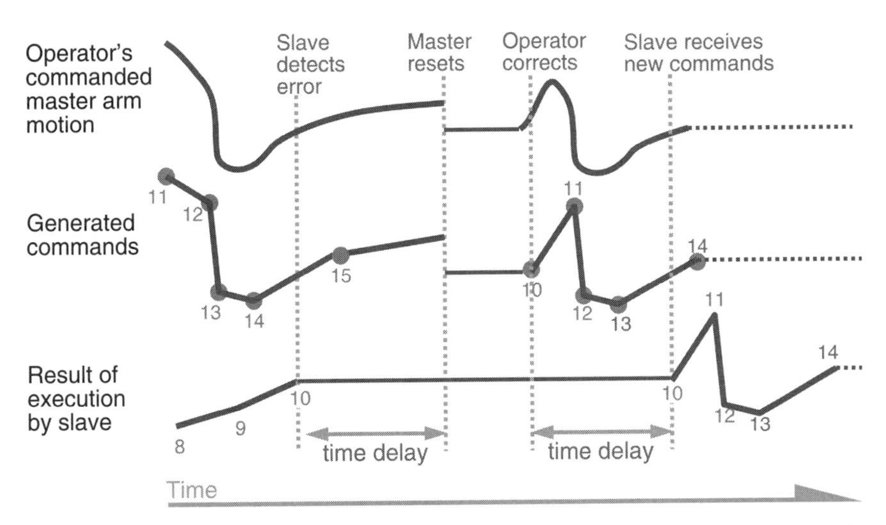

FIGURE 5.2: When the slave detects an error, it pauses and signals the operator station. Once that message reaches the master station it resets. The operator must then diagnose the problem to continue with the task.

Deciding what information to send between sites and when to send it is the most challenging and fundamental problem when performing teleoperation via constrained communications links.

5.4 Caches

One can think of the teleprogramming system as being analogous to a microprocessor cache. In a microprocessor, there is a fast internal processing engine connected to a large, but relatively slow, memory. To keep the processor fully active, portions of the memory most likely to be needed are copied into a small fast memory (a cache) close to the processor. The algorithms for deciding which parts of memory to place in the cache are based on usage and, occasionally, prior knowledge.

Think of the predictive display at the operator station as being a cache of information about the remote site, and think of the reflexes at the remote site as being a cache of information about how the operator would like the slave to react.

Neither "cache" contains a full copy of the whole real "memory", but if we choose carefully which information is stored in each, then just enough may be available to allow us to get the task done reliably without wasting bandwidth.

5.4.1 Prefetching

In a microprocessor, a cache prefetch is where the system attempts to predict what will be required and pulls it into the cache. For example, if you're viewing a small portion of a very wide image on the monitor then pieces of the image that are just out of sight can be prefetched so that, should you choose to pan the image left and right, they will be immediately available.

If the system's guess is correct, then we save the cost of a cache miss later. If its wrong, we consume memory bandwidth with the unnecessary cache load.

In the case of teleprogramming, it is possible to predict in advance which sensory feedback would be of most help to the operator in diagnosing an error (see Chapter 8). This information may be transmitted from slave to master so as to be immediately available should an error actually occur.

5.4.2 Cache miss

A cache miss is where the processor requires data that is not currently in the cache. This is an expensive operation. Everything stops, the data is copied from main memory into the cache, and processing continues.

This is analogous to an unexpected event at the remote site. The slave system requires knowledge of what to do, yet that is only available at the master station. Thus, the slave must pause, send a request to the master station, and await the arrival of new instructions.

At the operator station, the equivalent of a cache miss is a situation where the operator requires knowledge that is not included in the local model of the remote site.

In either case, a miss requires at least one round-trip communication between sites. That's expensive, so our goal is to eliminate as many cache misses as possible. We can do this by prefetching, but each prefetch is not without its own cost. Thus, we need to trade off between the certain, but relatively small, cost of a prefetch and the possible, but relatively large, cost of a cache miss.

5.4.3 Coherency

The cache maintains a copy of a portion of main memory. Thus, for some memory locations there are two copies, one in the cache and one in main memory. Now, if the processor changes the copy in the cache or some other device changes the contents of main memory, there will be two different copies of the same memory location. This is undesirable.

There are two common ways to avoid this. The first is a write-through cache, where any change to the cache is simultaneously propagated straight to main memory. The second is to watch for changes to main memory and invalidate those regions of the cache that are affected, forcing them to be reloaded if the processor needs them again.

For teleprogramming, the equivalent problem is maintaining a valid model at the operator station of the remote environment. Due to the communications delay it is impossible for the model to always be up to date. The best we can do is to detect and quickly react to cases where a discrepancy occurs. This is achieved through error detection and recovery (see Chapter 9).

5.4.4 Predictive branches

Another technique used in microprocessors to improve performance is the predictive branch. Here the system attempts to predict which way a branch will go so that it can begin to execute following commands before it has finished computing the decision on which the branch is based. If it is correct, then no time is wasted. If it is wrong, then all the work done on following commands must be discarded.

Now, we can't accurately predict what the operator will do next, but in rare cases we can predict a few expected possible outcomes of an action at the remote site. In such cases we can code responses for all expected outcomes and allow the slave to choose, at execution time, that path through the command tree that best fits the current conditions. Here we are predicating, at command generation time, the likely slave state.

5.5 Summary

The teleprogramming system combines a virtual reality representation of the remote environment with a low level of remote site intelligence. At the master station, the local model acts as a cache of information about the remote site. At the slave station the commands received from the master station provide sufficient knowledge to enable the slave to react immediately to sensory input.

The system observes the operator's interaction with a simulation of the remote environment and generates robot program instructions for transmission to the remote site. Should an error occur, the slave pauses and signals the operator who must then diagnose and correct the error, generating new commands for the remote robot.

6

A Natural Operator Interface

Now, return to the chair with a view of the sea from the Introduction. Imagine the slave robot is at some distant location, the TV is back at your feet, and the controller is in your hand. In presenting this example, we finessed away much of the difficulty. Let's take a closer look.

First, we assumed that when you moved your hand left, the robot on-screen moved left. This seems so obvious and simple. Yet it is not always true. In Section 3.4.2, prior work was described that indicated the need for performing this mapping in a natural manner.

Second, we assumed that it could duplicate your motions exactly. That would be true if the remote arm were a mechanical copy of your arm. If each of its joints has the same range of motion as yours and all the distances between its joints are the same as distances between joints in your arm, then it can exactly duplicate your motions. In this special case, it and your arm are considered to be *kinematically identical.*

Third, we assumed that the master arm (connected to the sphere in your hand) and the remote slave manipulator were always able to duplicate your motions. Unfortunately, having a mechanical arm with the same dexterity as a human arm often doesn't work out well in practice—the constraints when designing with steel and wires are quite different than those that have molded the evolution of arms constructed from bone and muscle.

Thus, well-engineered robot arms will tend not to be kinematically similar to human arms. Therefore, in mapping your motions to those of the remote robot, we must consider cases where the workspace of your human arm, the workspace of the master arm, and the workspace of the remote slave arm all differ.

In this chapter, the mechanisms used within the teleprogramming system for transforming operator input into motions of a slave robot are described. We'll consider how to perform this mapping in a natural manor, how to cope with cases where the master input device and slave output device are not kinematically identical, and look at some interesting cases where the input and output devices are very different.

6.1 The teleprogramming operator interface

In the teleprogramming system, the motion of the master input device (which the operator holds) is mapped into the virtual world (which the operator views) and used to control the motions of a simulated slave robot. The input device may be a conventional computer mouse or, more interestingly, a six-degree-of-freedom commercial robot arm (see Figure 6.1).

6.2 Creating a natural interface

We can define a "natural" interface as having the following characteristics:

- *Observability.* It should be possible to view the simulation from any desired viewpoint within the virtual world.
- *Continuity.* The simulation should appear to react instantaneously and should move smoothly.
- *Kinesthetic correspondence* [10]. The simulated slave should always appear to move in the same direction as the operator's hand.

6.2.1 *Observability*

Observability requires that the operator be able to change his or her viewpoint within the virtual world. This is a well-studied problem (see for example [26]). The present implementation adopts the conventional approach

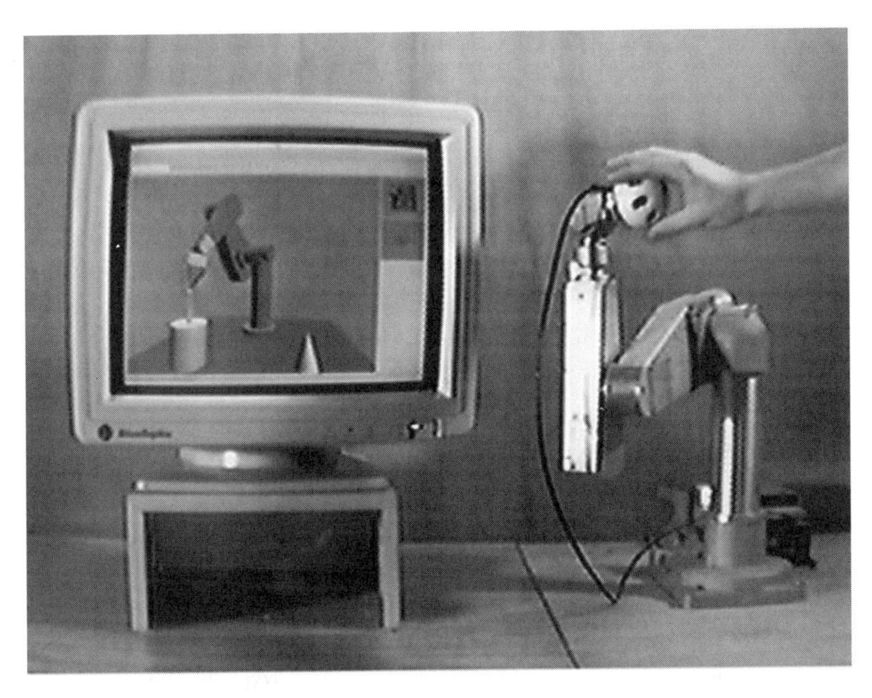

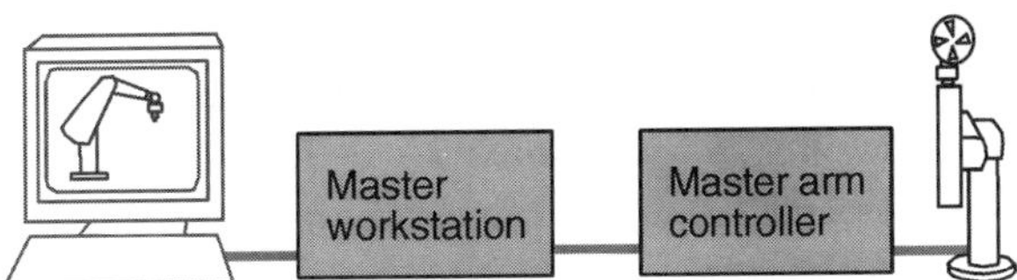

FIGURE 6.1: Teleprogramming operators view a graphical representation of the real remote slave site. In the depicted implementation the operator commands the simulated slave robot by holding and moving the end of the master arm. Our goal is to have the operator's interaction with the system seem natural.

of using a mouse to alter camera orientation and position. In keeping with the remainder of the system, the motion of the mouse while altering camera viewpoints is mapped into the virtual environment based on the operator's current viewpoint.

6.2.2 Continuity

Continuity is related to two factors: the rate at which the screen is redrawn, and the delay between operator action and observed system response. This is largely determined by the hardware employed. (For example, the frame redraw rate for our implementation varies between 3 and 60 frames/second depending on the particular graphics machine employed.) However, assuming that the hardware is fixed, there is a possible tradeoff between frame rate and lag. One option is to process a single frame at each instant. This is simplest and gives the smallest possible lag. An alternative is to process several frames in a pipelined fashion. This increases the average frame rate but also increases the lag. The author's intuition was that lag was more important than update rate and so, for the present implementation, the former option was employed. Recent work appears to support this view [95].

6.2.3 Kinesthetic correspondence

The implementation of kinesthetic correspondence requires that the operator's input motions (and any output force-feedback) are functions of the operator's chosen viewpoint within the virtual model. For example, if the operator moves the input device toward his or her left, then the simulated slave robot should appear to move toward the left (see Figure 6.2).

A sense of scale is also important. The operator must be able to move a, possibly huge, slave arm through its entire workspace while using a, probably relatively small, master input device. He or she should also be able to perform very precise delicate operations without requiring a very precise master device. By scaling the operator's actions based on their chosen viewpoint, we can allow him or her to choose between performing gross motions on a coarse scale or small motions on a fine scale (see Figure 6.2). The implementation is further complicated by computational constraints and the need to support reindexing.

Here we begin by considering the simplest case, where only translational motions are possible, and then work up to the full implementation, that

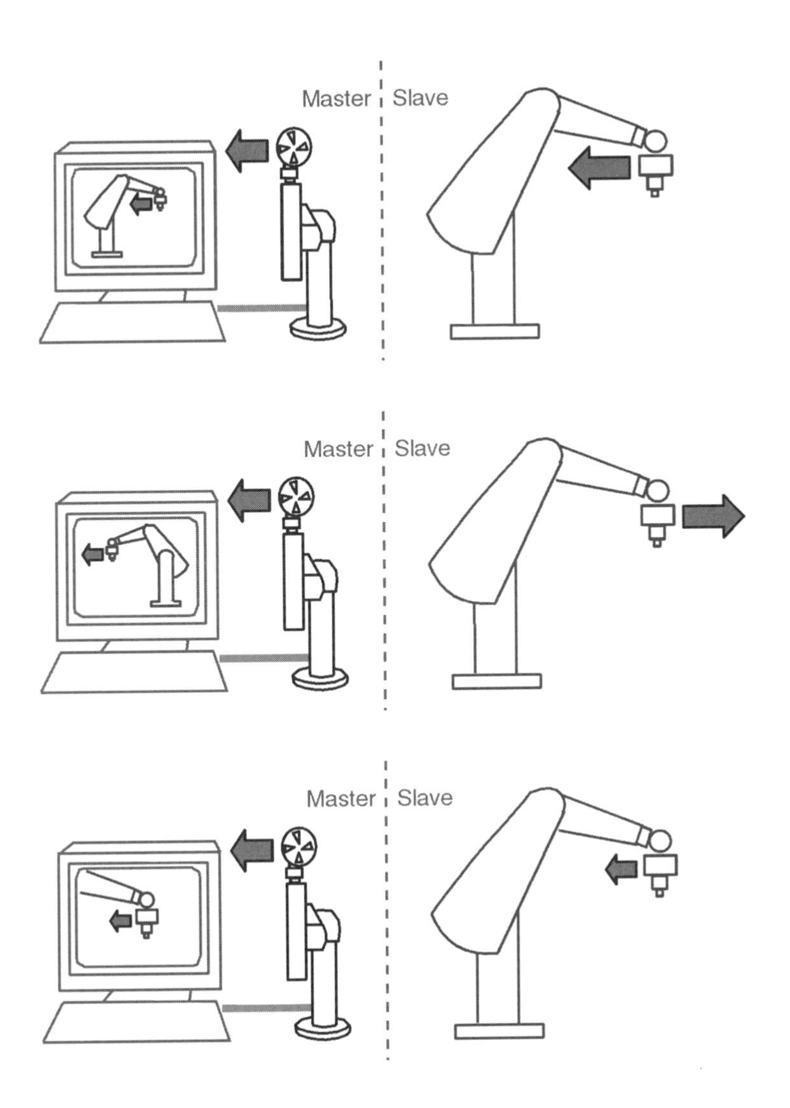

FIGURE 6.2: In the teleprogramming system, we want the operator's interaction with the system to appear very natural. Shown here are three examples of the same operator arm motion being translated into motion of the virtual slave robot on-screen and the corresponding motion of the real slave robot. Notice how the direction with which the operator views the slave robot changes the direction of the commanded motion (kinesthetic correspondence). Notice also how the operator may zoom in to perform small delicate operations (scaling).

allows arbitrary translations and rotations while mapping the operator's motions to maintain kinesthetic correspondence.

Definitions:

R^m The rotational matrix describing the orientation of the master manipulator end-effector. This is with respect to the master arm's kinematic base frame. A subscript of i indicates the initial position (just after reindexing), while a subscript of n indicates the current or new position.

$\mathbf{p}^m$ The position of the master end-effector relative to the master arm's kinematic base frame. Subscripts as above.

R^s The rotational matrix describing the orientation of the slave manipulator end-effector. This is with respect to the slave arm's kinematic base frame. Subscripts as above.

$\mathbf{p}^s$ The position of the slave end-effector relative to the slave arm's kinematic base frame. Subscripts as above.

s^{ms} The scale factor from master to slave coordinates.

$\mathbf{p}^{ms}$ The translation from master to slave coordinates.

R^{ms} The rotation from master to slave coordinates.

d The apparent distance of the operator from the simulated slave robot. In the current system, this is the distance within the virtual world between the slave kinematic base frame and the focal point for the camera.

k A constant scale factor between master and slave systems. For the present implementation, the value of k is selected so that, when the slave robot workspace is fully visible in the display, the slave may be moved from one side of its workspace to the other without any need to reindex the master arm.

Case 1: Only translational motions are permitted, and the master and graphical image are oriented so that the axes of their kinematic base frames are parallel.

$$s^{ms} = d \cdot k \tag{6.1}$$

$$R^{ms} = \text{identity} \tag{6.2}$$

$$\mathbf{p}^{ms} = \mathbf{p}_i^s - \mathbf{p}_i^m \cdot s^{ms} \quad (6.3)$$

$$R_n^s = R_i^s \quad (6.4)$$

$$\mathbf{p}_n^s = \mathbf{p}^{ms} + \mathbf{p}_n^m \cdot s^{ms} \quad (6.5)$$

Case 2: Both translational and rotational motions are permitted. The master and graphical image are oriented so that the axes of their kinematic base frames are parallel.

$$s^{ms} = d \cdot k \quad (6.6)$$

$$R^{ms} = (R_i^m)^{-1} \cdot R_i^s \quad (6.7)$$

$$\mathbf{p}^{ms} = \mathbf{p}_i^s - \mathbf{p}_i^m \cdot s^{ms} \quad (6.8)$$

$$R_n^s = R_n^m \cdot R^{ms} \quad (6.9)$$

$$\mathbf{p}_n^s = \mathbf{p}^{ms} + \mathbf{p}_n^m \cdot s^{ms} \quad (6.10)$$

Case 3: Both translational and rotational motions are permitted. The master and graphical image now have arbitrary orientation. Let C be a correctional rotational matrix obtained from the viewing matrix.

$$s^{ms} = d \cdot k \quad (6.11)$$

$$R^{ms} = (R_i^m)^{-1} \cdot (C \cdot R_i^s) \quad (6.12)$$

$$\mathbf{p}^{ms} = (C \cdot \mathbf{p}_i^s) - \mathbf{p}_i^m \cdot s^{ms} \quad (6.13)$$

$$R_n^{s'} = R_n^m \cdot R^{ms} \quad (6.14)$$

$$\mathbf{p}_n^{s'} = \mathbf{p}^{ms} + \mathbf{p}_n^m \cdot s^{ms} \quad (6.15)$$

$$R_n^s = C^{-1} \cdot R_n^{S'} \quad (6.16)$$

$$\mathbf{p}_n^s = C^{-1} \cdot \mathbf{p}_n^{S'} \quad (6.17)$$

This formulation was chosen with the intention that the calculations be distributed efficiently across processor boundaries (see Figure 6.3). Equations 6.11,6.12 and 6.13 are precomputed by the master station. Equations 6.14 and 6.15 are evaluated continuously during operation by the master arm controller, while equations 6.16 and 6.17 are performed continuously by the master station. This division ensures that the master arm controller, which has the strictest real-time requirement, performs the least computation. Note that, since the master arm position is being communicated in absolute coordinates (rather than as a relative motion), it is not necessary for the master station and master arm processors to be synchronized. The master station may simply use whatever transform has most recently been

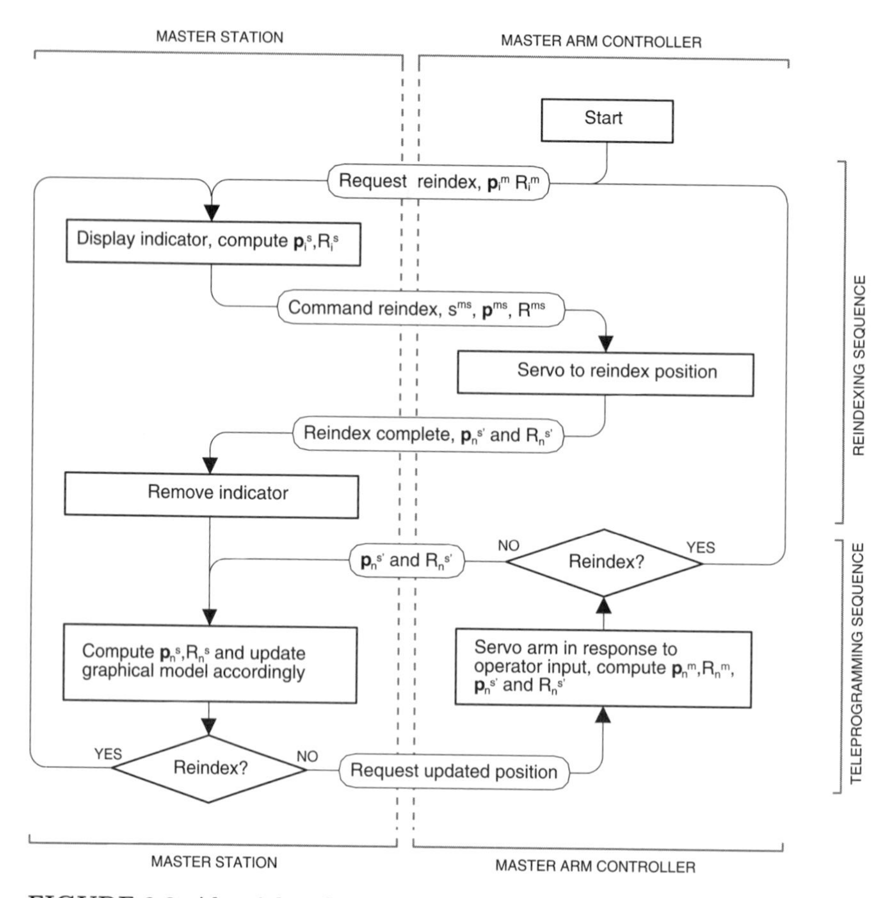

FIGURE 6.3: Algorithm for controlling the master input device for the teleprogramming system showing the flow of information between the master station and master arm processors. Note that, while teleprogramming, either processor may initiate a reindexing operation.

computed by the master arm controller. Note also that the values of s^{ms}, R^{ms}, and $\mathbf{p}^{ms}$ are dependent on the viewing position within the virtual world and must therefore be updated whenever this changes. In practice, this is implemented by reindexing the master arm whenever the operator changes his or her viewpoint within the virtual world.

6.3 The degree-of-freedom problem

In the simplest case, both the master and slave devices have six degrees of freedom. This results in a natural mapping between the master and the virtual world, and then between the virtual world and the simulated slave. There are, however, cases where this simple view is inapplicable. This may be permanent, as in the case where a five-DOF slave is employed, or temporary, as in the case where singularity or joint space considerations temporarily limit the available motion.

Let N_M and N_S be the number of degrees of freedom on both the master and slave devices. Assume that a "superior" device is one with more than six degrees of freedom, while an "inferior" device is one with fewer.

6.3.1 Inferior master device

In this situation, the master input device has insufficient degrees of freedom to permit simultaneous control of all Cartesian directions and orientations within the virtual world. The solution is to map the available degrees of freedom of the input device into the workspace of the simulated output device. By switching between a number of different mappings, it is possible to control all available degrees of freedom. Those actions that require simultaneous motion of more than N_M degrees of freedom cannot easily be performed with this approach. However, in practice, most desired motions may be successfully approximated. This is analogous to the situation in computer graphics applications where the operator must interact with a 3-D graphical model using only a 2-D input device.

The most common example of this mapping in the current implementation is when a mouse is used to command a six-DOF slave robot. This is dealt with by selecting a *plane of control* within the virtual world based on the operator's chosen viewing direction and then projecting the mouse motion onto that plane (see Figure 6.4).

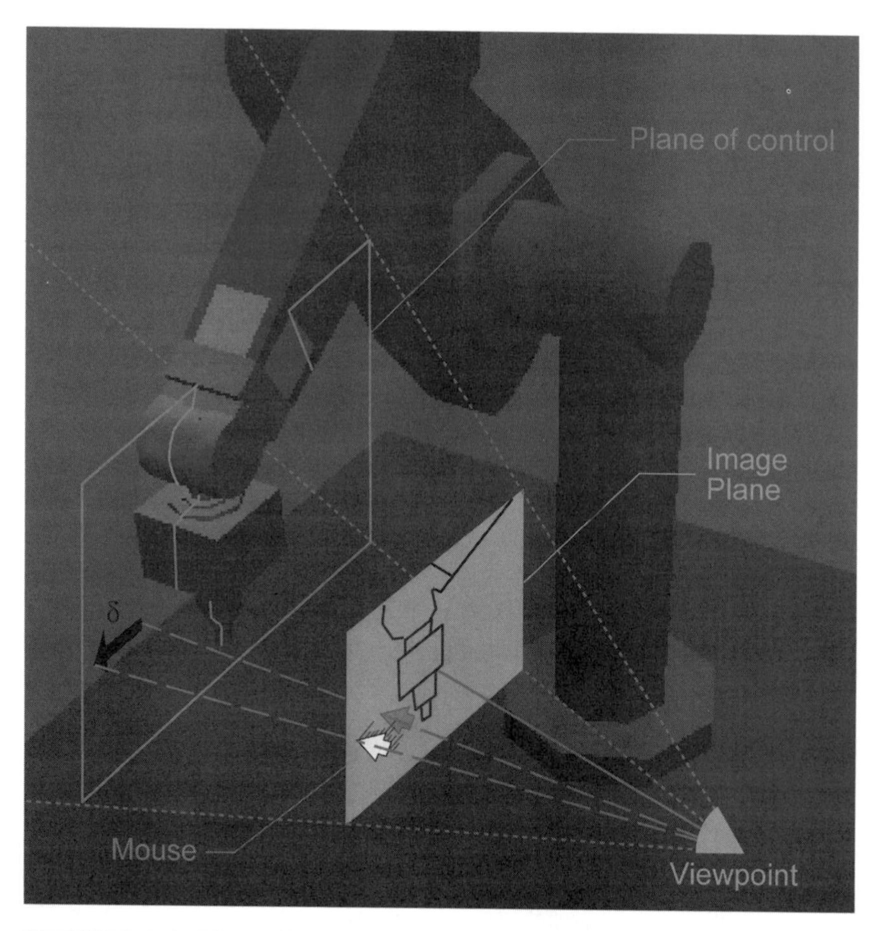

FIGURE 6.4: The 2-D motion of the mouse in the image plane is mapped onto the plane of control in the virtual world.

Definitions:

$\mathbf{v}_E$ is the position in the virtual world from which the operator is choosing to view the virtual world.

$\mathbf{n}_E$ is a normal representing the operator's chosen viewing direction within the virtual world.

$\mathbf{v}_C$ is that vertex on the remote robot that the operator wishes to control. Common examples are the intersection of the axes of rotation of the last two joints or the center of the end-effector.

$\mathbf{v}_M$ is the position of the mouse on the image plane in virtual world coordinates. Subscripts denote the initial and final mouse positions.

Then:

$$P_C = \text{plane through } \mathbf{v}_C \text{ with normal } \mathbf{n}_E \quad (6.18)$$
$$L_{Si} = \text{line through } \mathbf{v}_E \text{ and } \mathbf{v}_{Mi} \quad (6.19)$$
$$L_{Sf} = \text{line through } \mathbf{v}_E \text{ and } \mathbf{v}_{Mf} \quad (6.20)$$
$$\mathbf{v}_i = \text{intersection of } L_{Si} \text{ and } P_C \quad (6.21)$$
$$\mathbf{v}_f = \text{intersection of } L_{Sf} \text{ and } P_C \quad (6.22)$$

For translational motions, let δ be the 3-D vector by which the point of control will be translated within the virtual world (see Figure 6.4).

Then:

$$\delta = \mathbf{v}_f - \mathbf{v}_i \quad (6.23)$$

For rotational motions, let α be the angle by which the end effector of the robot being controlled should be rotated. The axis of rotation is that line which is parallel to $\mathbf{n}_E$ and passes through $\mathbf{v}_C$.

Then:

$$\mathbf{n}_i = \frac{\mathbf{v}_i - \mathbf{v}_C}{|\mathbf{v}_i - \mathbf{v}_C|} \quad (6.24)$$
$$\mathbf{n}_f = \frac{\mathbf{v}_f - \mathbf{v}_C}{|\mathbf{v}_f - \mathbf{v}_C|} \quad (6.25)$$
$$\alpha = cos^{-1}(\mathbf{n}_i \cdot \mathbf{n}_f) \quad (6.26)$$

Motion toward or away from $\mathbf{v}_C$ does not change the value of α but does change the precision to which it may be controlled. Rotations should be disabled when this distance becomes too small.

A slight refinement on the above is to rotate the plane of control to the nearest world coordinate axes, causing the axis of rotation to always be parallel to the world coordinate axes. This makes it easier for the operator, since he or she need not exactly duplicate prior viewing locations in order to duplicate, or negate, prior motions.

The system can assist the operator by automatically controlling those degrees of freedom that the input device cannot control. For example, in the current implementation when the operator performs translational motions, the system will attempt to keep the end effector orientation constant with respect to the robot kinematic base frame. Similarly, while the operator performs rotational motions, the system attempts to maintain the end-effector position. Synthetic fixtures (described later) provide a means to more actively control the passive degrees of freedom.

6.3.2 Superior master device

In this case, the master manipulator has more than six DOF. This is the classic redundant manipulator problem [98]. Six of the available degrees of freedom are constrained by the operator's grasp on the end-effector, and the system is free to control the remaining degrees of freedom. Within the teleprogramming system a superior master device is equivalent to one with only six DOF. The only benefit from additional degrees of freedom is that they may increase the available workspace, thereby making reindexing less frequent.

One interesting possibility is that, since the system has some knowledge of the task being performed, it can attempt to predict the operator's next desired action. It could then control the redundant degrees of freedom to maximize the probability that that action could be performed without any need to reindex.

6.3.3 Inferior slave device

In this case, the slave robot has fewer than six degrees of freedom. As a result, it will not always be able to perform those motions that the operator commands within the virtual world. This is a particularly undesirable situation since it often makes it impossible to achieve kinesthetic correspondence.

In cases where a master device with force feedback is employed, then the operator's motions may be constrained so as to discourage the operator from performing actions that are impossible at the slave site. An example of this is when a six-DOF master device is used to control the five-DOF JASON manipulator. The system could first calculate that rotation which would be necessary were the JASON arm to have a sixth joint, and then servo the master arm to null out that rotation.

6.3.4 Superior slave device

In this case, the slave manipulator has more than six degrees of freedom. As for the case of the superior master manipulator, this is the classic redundant manipulator control problem. In this instance, the end-effector position and orientation are controlled by the operator's motions within the virtual world, and the system is free to control the additional degrees of freedom.

The simplest technique is to selectively lock those joints that are not needed for the current Cartesian motion. This is the technique currently employed. A more refined method would be to use the task knowledge present at the master station to predict suitable positions for the redundant joints.

One particularly interesting situation is where the slave device is composed of several manipulators. In the current system, this is achieved by allowing the operator to switch between different slave robots, controlling first one, then another. When an error is detected, the system switches control to the affected robot and "rewinds time" for that manipulator to show the operator the point where the error occurred.

However, this approach does not provide any consideration for interactions between manipulators. If one arm were to fail, the other would continue to blindly execute subsequent commands, possibly compounding any error. To help avoid such difficulties, we may consider three different levels of coupling between remote manipulators: uncoupled, loosely coupled, and tightly coupled.

In the uncoupled case, there is no direct communication between remote manipulators, and thus, any dependencies must be enforced by the operator station. A possible solution is to buffer newly generated commands whenever the operator switches manipulators and not send those commands until the previously controlled manipulator has signaled that it has successfully completed all previously generated commands. However, this process only allows one manipulator to be active at any one time, and it does not solve problems that occur while manipulators are attempting to remain stationary.

We define a "loose" coupling between manipulators as being one with the resolution of a single teleprogramming command. That is, each manipulator knows the number of the command the other is executing and whether or not it was successfully completed. However, each manipulator does not know the exact position of the other. In this case, any error on one manipulator is known by every other one. These other manipulators may then also pause and ignore subsequent commands. This level of coupling would make it possible to add commands to the language to allow interdependencies to be explicitly stated. For example, we could add a "pause until robot x has completed command y" command.

A "tight" coupling, where each robot is aware of the other's position at servo rates, would make possible the coordinated control of multiple manipulators [100, 33, 87]. The teleprogramming interface could conceivably provide for very natural operator control of such a system—allowing him or her to control the position of either a real or imaginary object to which each manipulator was connected [38, 27, 74].

6.4 Summary

A natural interface requires observability, continuity, and kinesthetic correspondence. In the teleprogramming system the operator's motion of the master arm is interpreted, based on his or her chosen viewpoint within the simulated slave environment, to achieve natural motion.

The master and slave devices need not be kinematically similar. Examples were presented of operator control using both a small commercial robot arm (with six degrees of freedom) and a conventional computer mouse (with two degrees of freedom).

7

Synthetic Fixtures

Consider the problem of increasing the precision with which the operator may control the slave manipulator. One approach is to improve the visual information available to the operator. This could be achieved by using stereo systems [22], by providing additional visual clues with the addition of shadows and textures [93], and by superimposing visual enhancements onto the viewed scene [43]. A second, complimentary, approach is to improve the "feel" of the master arm by using improved hardware [52, 55]. The problem is that no matter how realistic the graphical simulation is, and no matter how perfect we make the master arm, we will still be limited by the accuracy of the human operator.

Consider, by way of example, the situation where an operator wishes to move the slave robot along a straight line in Cartesian space—as may be required during the task of mating two parts. The problem is that its very difficult for the operator to move the master arm in a perfectly straight line—even if the problem is reduced to only two dimensions it is still very hard. Consider trying to draw a straight line without using a ruler. One possible solution is to have the operator define starting and ending points and then have the system generate commands for motion between them. This could be accomplished with the "position clutch" described by Conway et al. [15]. However, the operator must still correctly specify the starting and ending points.

This is, in some sense, analogous to the problem faced by computer aided design (CAD) programs where its very difficult, for example, for a user to draw a line that exactly meets another line. In CAD systems, this is overcome by the use of a "pseudo pen location" [83] or "precision point assignment" [41] or, more recently, the "osnap" command [6]. These solutions to the precision problem all work by moving the cursor, not to exactly where the operator pointed, but rather to where the system thinks the operator intended.

What is needed for teleoperation is a similar feature that provides kinesthetic as well as visual feedback. We term this feature *synthetic fixturing* [70, 68, 69].

7.1 Overview

Synthetic fixtures provide the operator with task-dependent and context-sensitive visual and force clues. The system does not attempt to provide realism. Instead the intention is to provide the operator with those force and visual clues that can best aid him or her in task performance. For example, consider the case where the operator moves the end-effector toward a surface. If it is appropriate for the surface to be contacted, then the system activates an attractive fixture to assist the operator in moving to and then maintaining contact with the surface. Alternatively, if surface contact were inappropriate, then a repulsive fixture would be employed. The system does not just react to the operator's input; instead, it attempts to predict and then actively assist his or her actions.

Imagine the hypothetical example of a virtual reality simulation of a computer keyboard. If the intention were to give the user the illusion that her or she was using a particular real keyboard, then it would be important to accurately simulate forces. However, if our task is merely to help a user type, then it would be quite acceptable just to have every key feel approximately "key-like." Furthermore, by appropriate use of force clues, the operator could be provided with much more information than just whether or not a key was being pressed. For example, the force required to activate each key could be varied based on:

- Task knowledge—which keys should/should not be pressed at this time.
- Current configuration—which keys are currently being activated.

Thus, by providing force clues that are task- and configuration-dependent, the system can take a very active role in assisting the operator to complete a given task.

Synthetic fixtures are analogous to the "snap" family of commands used in computer aided design programs [83, 6, 9], increasing both the speed and precision with which an operator can work.

7.2 Operation

Fixturing operates in three stages. The first stage is activation where a fixture decides whether or not it should operate. These activation decisions should be made automatically—if operators had to manually select, activate, and de-activate fixtures, then much of the time saved by using fixtures would be lost. The system should attempt to predict the operator's expected actions and then, based on the task, activate those fixtures that can best assist the operator. A simple scheme is to have fixtures activate whenever an appropriate end-effector tool is nearer than some pre-specified distance.

If a fixture is active, then it provides both force and visual clues to the operator. During this second stage, visual clues assist the operator in deciding which actions to perform next, while force clues assist in actually performing those actions. These force clues are generated by perturbing the motion of the master arm. In the current implementation, the master manipulator is position controlled. It is servoed by reading the force the operator exerts (using a six-axis, wrist-mounted, force/torque sensor) and computing a new Cartesian set point for the arm motion based on those readings—interpreting them as the "desired" direction of motion. Fixturing is implemented by computing the distance from the end-effector to the fixture and then altering the set point as a function of that distance.

Fixtures are activated/deactivated when the end-effector is sufficiently far away as to apply no apparent force to the operator. This avoids problems with the end-effector appearing to jerk as fixtures are autonomously turned on and off.

If the operator allows the force clues to guide him or her to a position very close to some feature, then it is assumed that the operator intended to move exactly to that feature. The commanded slave position is then filtered to reflect this assumption. In this way, the system combines the operator's relatively imprecise motions with a little task knowledge to generate precise commands. This is analogous to the "snap" commands used in CAD programs. It has the effect of preventing positional errors at the operator station from "flowing through" to the rest of the system.

7.3 Terminology

We define most fixtures to be of type X–Y, where X is some feature on the end-effector and Y is some feature in the environment. Unless otherwise stated it is assumed that the system should aid the operator in bringing feature X on to feature Y. For example, a *point–point fixture* aids the operator in moving a point on the end-effector to a point in space, while a *closest-surface–surface fixture* assists in moving the closest surface (plane of finite extent) on a polygonal end-effector into contact with a particular surface in the world.

7.4 Command fixtures

Synthetic fixtures enable tools to be controlled in an intuitive manner without requiring additional physical controls. Consider the case where the end-effector is a tool that must be turned on and off. One possibility for controlling this tool would be to add physical controls such as push-buttons or foot-switches. However, it would be preferable if we could merely observe the operator's motions and then automatically generate commands to turn the tool on and off at appropriate times. This is complicated, since the operator's motions are often ambiguous. However, by using active force and visual clues, much of this ambiguity can be resolved. The basic idea is to place a virtual barrier around the position in which the tool should be used. By pressing through this barrier the operator clearly indicates his or her intention to activate the tool. The barrier is typically implemented as a closest-surface–surface fixture that calls a function whenever the end-effector penetrates it.

7.5 Example applications

In order to better describe the nature of fixtures, we present some specific examples ranging from simple single-fixture applications to more complex multi-fixture scenarios.

For these examples, the slave robot is a Puma 560 factory manipulator. The virtual remote site with which the operator interacts in depicted at left. This virtual world will initially contain synthetic cones and cylinders. These do not actually exist at the remote site. They are merely visual reminders of those places in the world with which the operator may wish to interact. In later examples, more realistic slave sites will be considered.

7.5.1 Point–point fixture

Here a *point–point fixture* is used to aid in bringing the tip of the end-effector (the dark cone) toward a point in the world (the tip of the thin cone). The operator has brought the end-effector sufficiently close for the fixture to activate, causing the bright star to appear on the screen. This visual indication of activation is important—it gives the operator some idea of what subsequent actions the system is expecting, and it typically precedes the use of any perceptible force clues. This gives the operator an opportunity to take corrective action if the system's "guess" as to his or her desired actions is incorrect.

At this distance, the operator feels a slight force pulling him or her toward the fixture point. This force increases as the end-effector is brought closer to the fixture. Typically, these force clues are sufficiently strong that the operator can feel and interact with them, but not so strong that motion in an alternative direction is prevented. In this way fixtures guide, but never completely constrain, the operator.

7.5.2 Point–path fixture

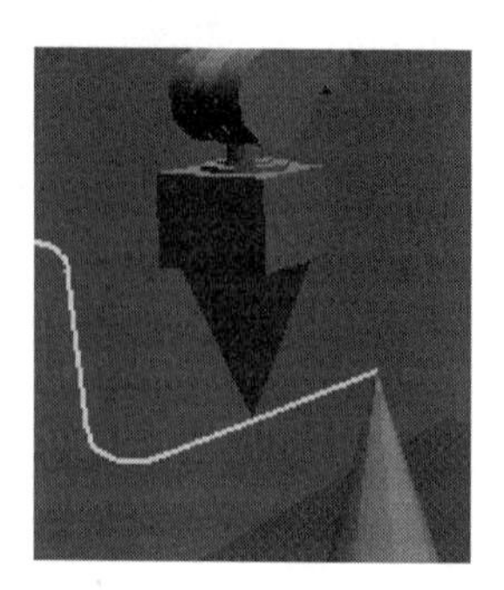

A *point–path fixture* is used to aid in bringing the tip of the end effector to a path composed of line segments and arcs. Moving the end-effector so as to trace along this path requires no effort, while moving away from the path requires that the operator overcome a distinct resistance. In this example, the operator is able to accurately trace along a 3-D path despite the fact that he or she may only be working with a 2-D visual display. The limited depth information available to the operator from visual means is supplemented by that available through force clues. They provide the operator with information that may be either unavailable or difficult to perceive from other sources.

In common with other fixtures, the path fixture was deliberately designed with the intention of allowing efficient rendering on-screen and efficient computation of distance for evaluation of the fixturing algorithms. Since only arcs and lines are employed, we may make use of existing graphics primitives for rendering these features, while simple closed-form solutions exist for determining the closest distance between a point and either primitive.

7.5.3 Closest-surface–surface fixture

In this example, a *closest-surface–surface fixture* is used to aid in bringing the end-effector (the dark box) into a face–face contact with a surface in the world. The operator has brought the end-effector sufficiently close for the fixture to activate, causing the surface to be highlighted (the hashed lines). At this point, the operator feels a slight force pulling the end-effector down and a slight moment rotating it toward a flat face-face contact.

If the operator decides to move closer to the surface, then the magnitude of these clues will increase, guiding the end-effector to a perfect contact.

Surface fixtures are not an attempt to simulate forces which the operator might feel if he or she were touching a real surface. In particular, they provide force clues before contact occurs. Rather than just telling the operator when he or she has hit a surface, they actively guide the end-effector toward flat face-face contacts. The feeling for the operator is similar to moving a magnetized block around in a frictionless metal world. The force clues are strongest when the end-effector is very close to a face-face contact and decrease as the end-effector moves away from this position. If the operator chooses to push hard enough to force the end-effector through the surface, then the force clues will decrease the further the effector is moved inside. Thus, regardless of whether or not the operator chooses to comply with a fixture, its effects are only felt while in close proximity.

7.5.4 Multiple fixtures for box interaction

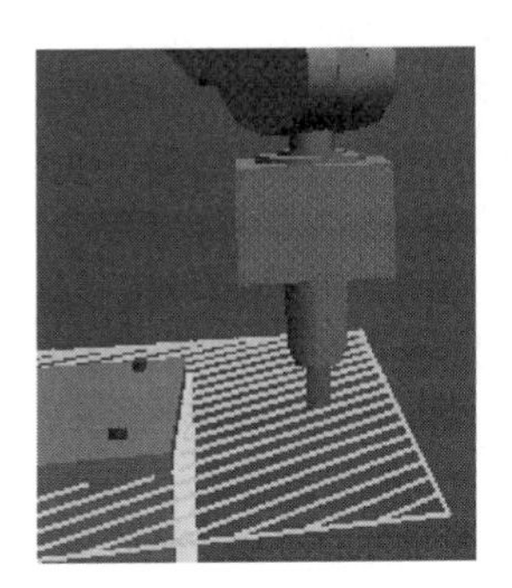

In this example, the box end-effector has been replaced with an impact wrench, and the simple geometric shapes have been replaced by a model of a real wooden box. The operator's task is to unscrew the bolts holding down the cover of that box. The first step in accomplishing that task is to contact the box in several places to refine the modeled location of the bolts. Here a *surface–surface fixture* is used to aid the operator in keeping the bottom cylindrical portion of the impact wrench aligned approximately 1 cm below the top surface of the box at left of the picture. Moving in the plane of the surface requires no effort, while moving either up or down requires a deliberate effort. This fixture provides an imaginary plane on which the operator may slide in order to contact the side of the box with the socket. Without the fixture, it is difficult for operators to maintain just the right height—if too high then the socket will miss the box, while if too low the upper portion of the impact wrench will contact the box. Note that a second surface fixture has activated on the side of the box as the system anticipates the operator's next motion into contact.

7.5.5 *Multiple fixtures for bolting/unbolting*

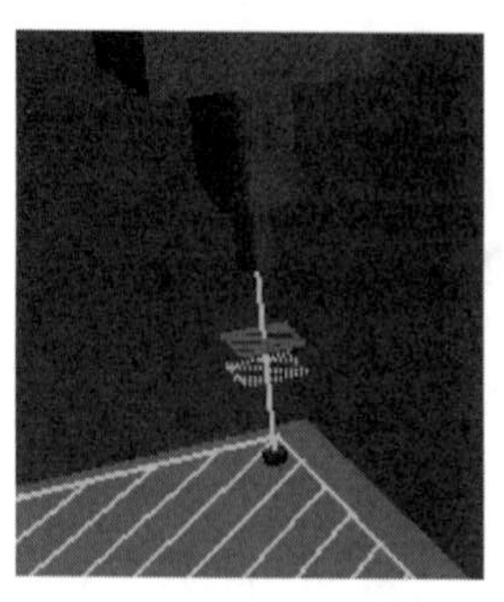

Here a number of fixtures are used to assist the operator in performing a bolting/unbolting task using the impact wrench. A *point–line fixture* has aided the operator in bringing the end-effector to a point where it is vertically over the bolt. The two square fixtures between the end-effector and the bolt form a virtual pushbutton. The upper surface (a *command surface–surface fixture*) acts as the top surface of the button, while the lower one (a *repulsive surface–surface fixture*) acts as a detent limit. By pressing the impact wrench down through the upper surface, the operator commands the system to activate the impact wrench.

All three of these fixtures are attached to the bolt hole so they will move with it if the world model is altered. They have activated in this case only because the end-effector is an impact wrench, it has a socket of the appropriate size, and the bolt hole contains a bolt. They would not have activated otherwise. A surface fixture on the top face of the box has also been activated. This is because the system cannot be certain whether the operator intends to slide down the line fixture to activate the impact wrench, or alternatively whether he or she intends to move away from the bolt and touch the surface with the empty socket. In cases such as this, the system activates several fixtures and allows operators to choose, by their actions, those which are most appropriate.

7.6 Algorithm

The previous sections introduced the concept of synthetic fixturing. However, if fixturing is to be practical, then an efficient algorithm is required. This is particularly important since the activation decisions should be computed at the frame update rate for the screen, while the force clues due to active fixtures must be recomputed at the rate of the Cartesian servo loop for the operator's hand controller. We begin by looking at one of the simplest fixtures and then work up through increasingly more complex implementations. For simplicity throughout this section it will be assumed that everything is in the same coordinate frame. In practice, this is unlikely to be the case—particularly since it is desirable to maintain kinesthetic correspondence (see Chapter 6).

7.6.1 Point-point fixture

Let e be a label assigned to the location on the end-effector that is to be brought in contact with the fixture. Let $\mathbf{p}(e)$ be the initial position of e, and let $\mathbf{f}$ be the position of the fixture point.

Let $\mathbf{h}(e)$ be a vector from the fixture to the end-effector point.

Then:

$$\mathbf{h}(e) = \mathbf{p}(e) - \mathbf{f} \tag{7.1}$$

Now, consider a simple activation function, choose a constant distance T_a and activate the fixture whenever the end-effector point is inside a sphere of radius T_a centered on $\mathbf{f}$, i.e. activate whenever:

$$|\mathbf{h}(e)| < T_a \tag{7.2}$$

Generate force clues by perturbing the end-effector to bring e closer to $\mathbf{f}$ using the motion perturbation function:

$$\mathbf{p}'(e) = \begin{cases} \mathbf{p}(e) - \mathbf{h}(e)W(|\mathbf{h}(\mathbf{e})|) & \text{if } |\mathbf{h}(e)| < T_a \\ \mathbf{p}(e) & \text{otherwise} \end{cases} \tag{7.3}$$

where $\mathbf{p}'(e)$ is the location of the point on the end-effector after perturbation by the fixture, and $W(d)$ is the fixture weighting function. A typical example being:

$$W(d) = \frac{1}{1 + k_1 d^{k_2}} \tag{7.4}$$

where k_1 and k_2 are constants determining the "feel" of the fixture.

Then filter the commanded end-effector positions:

$$\mathbf{p}''(e) = \begin{cases} f & \text{if } |\mathbf{h}(e)| < T_f \\ \mathbf{p}(e) & \text{otherwise} \end{cases} \tag{7.5}$$

where $\mathbf{p}''(e)$ is the filtered position of the end-effector point, and T_f is the filtering tolerance.

In crafting a point-point fixture, there are four constants to choose. T_a determines the activation distance. It is a compromise between a large value, which would make it possible for many fixtures to be simultaneously active, and a small value which would limit both the system's ability to guide operator motion and the operator's ability to react to inappropriate fixture activation. T_f controls the distance at which filtering is effective. Too large a value will unnecessarily constrain commanded actions (since any operator-chosen position within T_f of f will be commanded as though it were exactly f); while too small a value will allow any imprecision in the position of the master arm to influence the precision to which the slave manipulator may be commanded. The two weighting function constants, k_1 and k_2, determine the "feel" of the fixture. Their choice is influenced by operator preference, the desirability of having large force clues within T_f of the fixture, and the necessity of having imperceptibly small force clues at a distance of T_a from the fixture. Typical values are $T_a = 50\ mm$, $T_d = 0.1\ mm$, $k_1 = 0.01$, and $k_2 = 2$.

Now, this simple activation will work, provided that it is always appropriate for that end-effector point to be drawn toward that location in the world. In practice, this is unlikely to be the case, so the activation function will often be predicated on the current state of the world model and the current end-effector. In addition, more complex functions will sometimes be necessary, particularly if there are obstacles between the end-effector and the fixture.

7.6.2 Point-path fixture

Having examined the point fixture, we can now build on that to create more complex fixture implementations. Consider the point-path fixture. In choosing a representation for the path, we need a method that provides both rapid rendering and rapid distance computation. A piecewise linear path would be simple to render, but it's undesirable since the intersection points are not smooth. A series of spline curves could have been employed, but distance computations for such curves are less efficient. So, the chosen representation was a sequence of arcs and line segments. These paths may be rendered using existing graphics primitives, and the distance computation is rapid since simple closed-form solutions exist for both.

Fixture operation involves first finding the closest point on the path to the end-effector point and then applying the same algorithms as for the point-point fixture. This has the desirable side effect that the ends of the paths are handled in a natural way.

7.6.3 Closest-surface–surface fixture

In common with other fixtures, the closest-surface–surface fixture requires an activation function, a motion perturbation function, and a filtering function. For now, assume that the end-effector is a convex polyhedral shape and that the fixture surface is a plane of infinite extent (the case of a finite plane will be considered later). Define the fixture plane as having normal $\mathbf{v}$ and minimum distance to the origin of k. Let $\mathbf{p}(c)$ be the position of the end-effector vertex labeled c. Then the distance of a vertex to the surface is:

$$d(c) = \mathbf{v}.\mathbf{p}(c) - k \tag{7.6}$$

For the moment, assume a simple activation function where the fixture is active whenever

$$-T_a < \min(d(c)) < T_a \tag{7.7}$$

where T_a is some pre-specified activation tolerance.

Once the fixture has activated, end-effector motion perturbation is computed as follows:

1. Find the three vertices that have minimal distance to the plane and label them c_1, c_2, and c_3, where $d(c_1) \leq d(c_2) \leq d(c_3)$. It is assumed that these three points all lie on the end-effector face which is closest to the surface—this is not a general solution but it works well for many convex polyhedral end-effector shapes.

2. Move the end-effector set-point to bring c_1 closer to the surface:

$$\mathbf{p}'(c_1) = \mathbf{p}(c_1) - \mathbf{v}d(c_1) \times \begin{cases} W_1(d(c_1)) & d(c_1) < 0 \\ W_2(d(c_1)) & d(c_1) \geq 0 \end{cases} \tag{7.8}$$

   Note that different weighting functions are used depending on whether or not the end-effector intersects the surface ($d(c_1) < 0$). Thus, the force felt by operators as they push through the surface can be made larger than that felt as they pull the end-effector away from the surface. Note also that if the end-effector is well "inside" the surface, then the force will decrease allowing the fixture to be deactivated without the operator feeling a significant change in force feedback.

3. Now rotate about a line that passes through c_1 and is normal to both $(c_1 - c_2)$ and $\mathbf{v}$ so as to bring c_2 closer to the plane with normal $\mathbf{v}$

that passes through c_1. If α is the angle to bring c_2 into contact with this plane, then rotate by an angle:

$$\alpha' = \alpha W_3(d(c_2)) \tag{7.9}$$

4. Now rotate about the line through c_1 and c_2 to bring c_3 closer to the plane with normal $\mathbf{v}$ that passes through c_1. If β is the angle to bring c_3 into contact with this plane, then rotate by an angle:

$$\beta' = \beta W_4(d(c_3)) \tag{7.10}$$

It would have been possible to achieve a similar effect by replacing the two rotations with a single rotation about a suitable axis through c_1, however, this would not give as much information to the operator. In particular, it would make it difficult for the operator to distinguish and maintain a line-face contact between the end-effector and the surface.

The filtering function is then computed as follows (where T_p and T_r are the positional and rotational tolerances of the system, respectively):

1. If $d(c_1) < T_p$, then generate commands under the assumption that the end-effector has at least one point in contact with the surface ($d(c_1) = 0$).
2. If $d(c_1) < T_p$ and $\alpha < T_r$, then assume the end-effector is at least in a line-face contact ($d(c_1) = d(c_2) = 0$).
3. If $d(c_1) < T_p$ and $\beta < T_r$, then assume the end-effector is in face-face contact with the surface ($d(c_1) = d(c_2) = d(c_3) = 0$).

In common with most fixtures, this filtering operation is equivalent to the motion perturbation function but with all weighting functions set at unity.

7.6.4 *Multiple fixtures*

When moving beyond simple examples toward realistic implementations it must be expected that multiple fixtures will simultaneously be active. This will occur because:

- When building complex constraints, it is preferable to combine several simple generic fixtures than to build a single complex fixture specific to each application.

- Ambiguous operator motions can cause the system to be uncertain as to which fixture should be activated next. A simple, powerful, solution to this is to activate several fixtures simultaneously and then allow the operator to choose, by their subsequent motions, those which are most appropriate.

So, in dealing with multiple fixtures, the system must cope both with fixtures that compete as well as with those that are complimentary. A possible solution is to treat the fixtures as being somewhat analogous to forces pulling the end-effector in different directions—computing the resulting force by simply adding the perturbations from all active fixtures. However, while this approach appears elegant it does not work well in practice. Consider a case where the end-effector is between two surfaces and the system is unsure which of those two surfaces the operator wishes to contact. If motion perturbations are summed, then it there is no stable position that is exactly in contact with either surface.

An alternative solution (and the one currently employed) is to apply motion perturbations sequentially to the end-effector with the sequencing chosen based on inverse effect—applying the fixture with the least influence first and those with greatest influence last. The result is that the end-effector may be brought exactly in contact with a close fixture even if a far fixture is simultaneously active. This means that the fixture that would have greatest effect (and hence, presumably the one whose effect is most important) will have the most control over the end-effector position.

7.6.5 *Fixture activation considerations*

The fixture activation functions described above have been predicated only on the current state of the world model. While appropriate for simple situations, this does not cope well with the general case where the system may be started with the end-effector at any position within the model. For example, consider the case where the system is started with a point on the end-effector very close to a surface fixture. In this situation, the fixture would be activated immediately and comparatively large forces would suddenly be applied to the operator's hand controller. This is clearly undesirable.

The solution is to make fixture activation dependent on the motion of the end-effector rather than on its current position. A fixture may activate, changing state from *off* to *on*, only if the end-effector is moving from outside the activation volume to inside the activation volume. It may deactivate whenever the end-effector is outside the activation volume.

For some fixtures, it is appropriate to further constrain activation to those cases where end-effector motion is in a particular direction. For example, imagine that the operator has pushed the end-effector through a surface fixture and that the fixture has deactivated. In such a situation, the operator has clearly indicated that he or she did not wish to be constrained by the surface and it would therefore be inappropriate for that fixture to reactivate if the operator chose to pull the end-effector back through that same fixture. To deal with these situations, we introduce the notion of an arming volume. Fixture activation requires that the end-effector pass from the arming volume to the activation volume. For simple fixtures, such as the point-point fixture, the arming volume is the entire world. Thus, any motion into the activation volume will cause the fixture to turn on. For the surface fixtures, we choose the arming volume to be that region for which $d(c) \geq T_a$. Thus, these fixtures activate if the end-effector is moving down from above, but not if it's moving up from within. Note that the arming and activation volumes may be six-dimensional, considering end-effector orientation as well as position.

The need for hysteresis in the fixture activation functions is particularly important for teleprogramming. Without it, infinite loops can develop. For example, consider the case where the slave detects an error, the operator station model is reset with the slave close to a fixture, the system restarts, the fixture activates and pulls the end-effector toward the fixture. This motion generates a new command for the slave robot, which causes another error, which again causes the operator station model to be reset ...

Fixture activation should further be predicated based on task knowledge and the configuration of the world model. In the current implementation, we achieve this by creating the virtual world from a series of intelligent objects. Each object observes the state of the end-effector and its own internal state to decide whether or not to enable activation of fixtures under its control. For example, the bolt hole object controls a line fixture (to assist the operator in moving the end-effector down an approach trajectory) and a command fixture (to activate "remove bolt from hole" or "insert bolt into hole" command streams). It enables these fixtures only if the following constraints are satisfied:

- The end-effector is an impact wrench, *and*
- The wrench has an appropriately sized socket, *and*
  - The socket is empty but the hole contains a bolt, *or*
  - The socket contains a bolt and the hole is empty

Now, imagine that the operator moves toward a fixture but it fails to activate. This could be because the operator's motion was inappropriate (perhaps they he or she is approaching the fixture from the wrong direction), or it may be that fixture activation was inappropriate (perhaps the wrong end-effector is in use). An additional visual clue is employed to allow operators to disambiguate these two situations. If the end-effector follows an appropriate trajectory, then the visual clue for the fixture is always activated but its color is altered depending on whether the fixture is enabled (and will provide force clues and filter effector motion) or disabled (and have no effect on the end-effector).

7.7 Application to teleprogramming

In the teleprogramming system, fixtures assist operators in moving quickly and precisely. The visual clues aid operators in deciding which action to perform within the simulated remote environment, while the force clues actively assist them in performing the chosen motions. In this respect, there is little difference between teleprogramming and pure virtual reality applications.

The distinguishing feature of teleprogramming is that the operator's actions within the virtual world are continuously observed and autonomously translated into commands for transmission to the real slave robot. Since operator motion causes motion of a real slave robot, the consequences for incorrect motion are much more severe than in the case of virtual reality—while a simulation may simply be reset to its previous state, the same is regrettably not true for the real world.

The process of translating operator actions into discrete commands is complicated by ambiguous motions. For example, if the operator moved close to, but not exactly in contact with, a surface, then the system must guess whether or not the operator really intended the slave to contact that surface. Such situations can be avoided by appropriate use of synthetic fixtures.

The visual clues due to active fixtures give the operator some indication of what commanded actions the system is capable of performing, while the force clues assist the operator in making very deliberate motions from which it is much easier to determine intent. In this way, fixtures provide an additional channel of communication between human and machine. They reduce the operator's uncertainty as to what command the system will generate while simultaneously reducing the system's uncertainty as to which command the operator wishes generated.

The nature of the generated teleprogramming commands means that the slave robot may reliably be commanded to come into contact with, and slide on, a surface. However, contact motions near the edge of a face are unlikely to work correctly in the presence of uncertainty. Thus, the system should should act to attract the end-effector near the center of a face, while repelling it when near an uncertain edge. This is achieved by combining two complimentary force clues and transitioning between them based on the location of the end-effector. When the end-effector is in contact with a portion of the face that is likely to exist at the remote site, then the force clues are generated as for the surface fixtures described previously. When near an uncertain edge, the end-effector is repelled from the surface.

Visual clues are used to provide the operator with information on both fixture activation and the uncertainty associated with the face boundaries. When activated, each face displays a green "hashed" region showing that region of the object on which contact operations may be performed reliably. The uncertain edges are marked by red lines. An example motion sequence is shown in Figure 7.1.

The use of fixtures to resolve ambiguity is most clearly seen in the use of command fixtures. For the teleprogramming implementation these fixtures provide a simple, powerful, and very natural way for the operator to command the system to perform preprogrammed action sequences.

7.8 Application to conventional teleoperation

In conventional teleoperation, there is typically no world model and, even if one existed, the low-level nature of the position commands transmitted to the slave site would make it difficult to cope with modeling uncertainty. This makes it difficult to apply synthetic fixtures to this application.

However, for at least some cases it may be possible to utilize a restricted form of fixturing even if a world model doesn't exist. For example, the operator could press a key to activate a "remember this point" action where both the master and slave stations would record their current positions. The master station could then guide the operator in moving back to that point by activating a point fixture at its location. If the operator chooses to comply with the fixture, then the motion command transmitted to the slave will be exact and the accuracy with which the slave moves back to that point will depend only on its positional accuracy.

More complex situations should also be feasible. The system could observe the operator and then activate fixtures to aid the operator in repeating or reversing some portion of the task. For example, if the operator activated an impact wrench, the system could autonomously create a fixture to assist the operator in moving along a line through the axis of rotation of the wrench.

However, the system really needs some knowledge of both the world and the task if it is to reliably anticipate the operator's actions. This type of force feedback is thus better suited to teleprogramming or virtual reality applications rather than conventional teleoperation.

7.9 Application to virtual reality

Synthetic fixtures are applicable to cases where the performance of a task takes precedence over providing "exact" force feedback. Examples of appropriate tasks include selecting among alternative options or moving the operator's viewpoint on the virtual world.

For example, one can envisage an application where a set of data is displayed as a set of nodes. The operator could navigate along different paths through those nodes with assistance from synthetic fixtures. One interesting possibility would be to vary the force required to move to a new node in proportion to the time required in order to retrieve and/or compute it. This would allow operators to "feel" their way through the data while compromising between the amount of data displayed and the time required to obtain it.

Another application would be controlling the position of the virtual cameras through which the operator views the virtual world. For example, we could

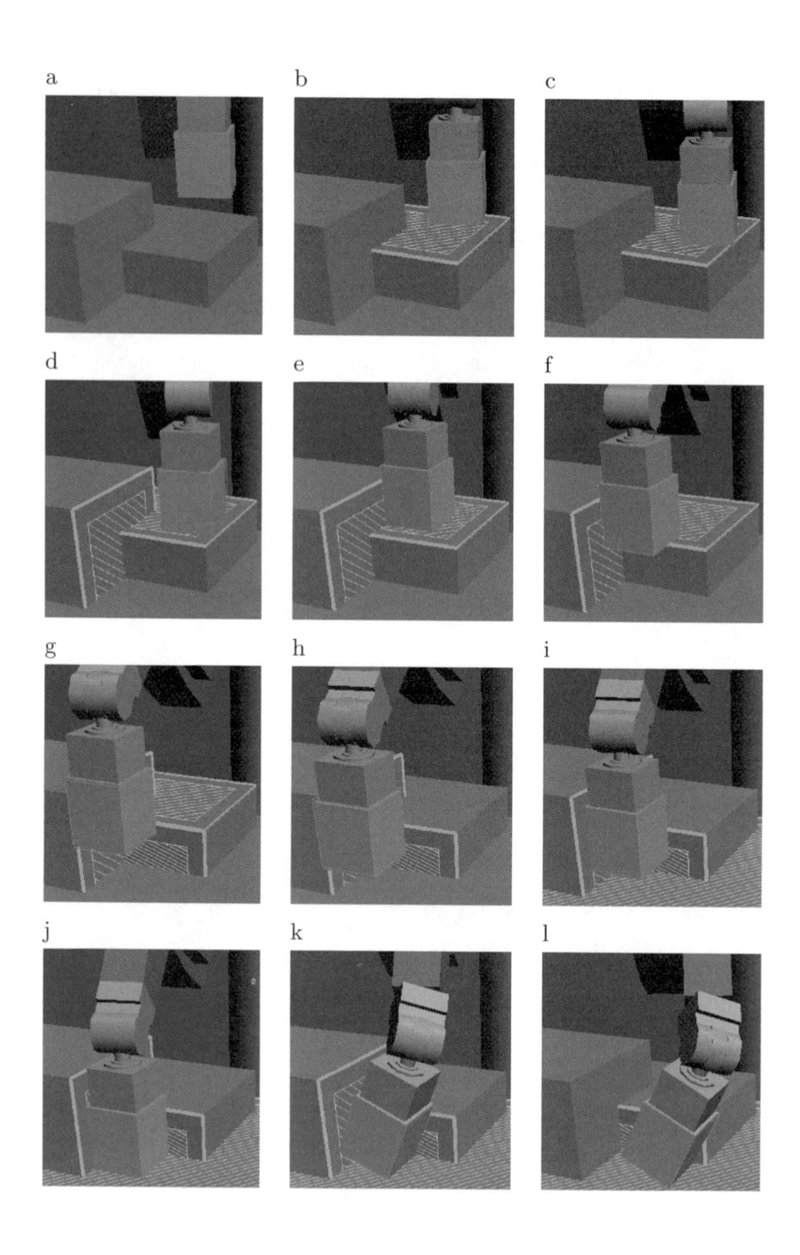
a
b
c
d
e
f
g
h
i
j
k
l

FIGURE 7.1: In this example, a number of fixtures are used to guide the operator as he or she commands a Puma 560 remote manipulator. The remote environment consists of several boxes of uncertain geometry. The end-effector is a box of known geometry. Initially, the end-effector is in free-space and there are no active fixtures (a). As the end-effector moves closer to the shorter box (b), a face-fixture is activated. This guides the operator, through force clues, to a face-face contact (c). At this point motion in the plane of the active fixture is still unrestricted but any other motion, such as lifting off the surface, requires a deliberate effort on the part of the operator. As the end-effector is moved closer to the taller box a second face fixture is activated (d), and the operator's motions are guided to a second face-face contact (e). Any motion other than that parallel to the intersection of the two active fixtures now requires a deliberate effort. As the operator attempts to slide off the top of the shorter box (f), the end-effector leaves the "certain" region of one face and that fixture becomes repulsive, pushing the end-effector up and away from the uncertain edge of the object. As the operator moves the end-effector forward and down, a third fixture activates (g). At this point the system is assisting the operator in maintaining a face-face contact with the tall box while dissuading him or her from moving near the uncertain edge of the shorter box. As the end-effector is moved down, the top face fixture deactivates (h) and the end-effector is guided into a second face-face contact (i). Yet another face fixture is then activated, and the end-effector is guided to a third face-face contact (j). At this point motion in any direction requires a deliberate effort on the part of the operator. Fixtures guide, but do not completely constrain, the operator. For example, the operator could still choose to rotate (k) and slide (l) away from active fixtures.

use positional fixtures to aid the operator in moving a camera among predefined locations.

In these cases there is no "real force" and it is appropriate to choose those force clues that best help the operator perform the task.

7.10 Alternative input devices

The preceding discussion of fixturing assumed the use of a six-DOF, force-reflecting, master input device. The most commonly suggested alternative to such an interface is to use a conventional computer mouse. In this case there are a number of tradeoffs.

The advantages of using a mouse are:

- Cost. A mouse costs approximately two to three orders of magnitude less than current force-reflecting hand controllers.
- Size. The mouse on the computer used to write this book requires a working volume of less than 0.01 m^3, while the current master manipulator requires 0.5 m^3.
- Safety. Being a lightweight and completely passive device, the mouse is unlikely to cause physical harm to the operator.
- Comfort. Using a mouse requires considerably less effort than a master manipulator. The weight of the operator's arm is supported by the table, motions are typically smaller, and he or she need never overcome inappropriate force feedback.

The disadvantages are:

- Limited dexterity. Having only two degrees of freedom, the mouse limits the ability of the operator to perform smooth motions in more than two DOF (see Section 6.3.1).
- Passivity. The mouse is unable to provide force feedback; thus, the system is unable to actively guide the operator. The ability to use force clues to inform the operator of desirable/undesirable motions is removed.

The optimal choice of input device will depend on the particular task, as well as on the preferences of the particular human performing that task. If the required operation maps well into a series of two-DOF actions, then using a mouse is preferable. Alternatively, if the task requires dextrous multi-degree-of-freedom motions, then one either has to use indirect methods to enter the trajectory specification [97], or one must use a more sophisticated input device.

Based on our experience with the teleprogramming system, a good compromise between cost and performance is likely to be provided by systems with two or three DOF and force feedback. Examples of such devices are the force feedback joystick [73], the planar haptic interface [23] and the PHANToM force-reflecting haptic interface [75].

The use of input devices with more than two or three degrees of freedom must ultimately be seen as an intermediary step. As systems become more sophisticated and are imbued with greater levels of task knowledge then the need for the operator to explicitly specify all input parameters must decrease. It will be sufficient for the operator to merely provide the system with sufficient hints so as to enable it to disambiguate between those different possible actions for which it has already planned full six (or more) degree-of-freedom motions.

7.11 Summary

Synthetic fixtures provide an operator with task-dependent and context-sensitive visual and force clues. Analogous to the snap commands used in computer drawing programs, they increase both the speed and precision with which operators may work.

Fixtures are applicable to teleprogramming and to those virtual reality applications where the timely performance of a task takes precedence over realism in force feedback.

8

Visual Imagery

This chapter describes the integration of remote site cameras into the teleprogramming system. Images from these cameras may be used for calibrating and maintaining the world model. However, the primary purpose is for improving the operator's "situational awareness"—providing him or her with the necessary information for diagnosing and resolving unexpected events at the remote site.

Since the communications bandwidth is limited, its infeasible to send every pixel from every camera back to the master station. Thus, we must trade off between the "value" of the information gained from an image fragment, and the "cost" in time and/or energy in obtaining and transmitting that information back to the operator.

8.1 Camera calibration

As with the calibration of most sensors, camera calibration involves the three steps of data gathering, interpretation, and computation. During data gathering, one or more images are taken with the camera. Interpretation requires that the raw pixel information from these images be converted into a symbolic form in which the relationship between pixels in the imaging plane and locations in the real world are characterized. Computation then involves determining calibration parameters from those known symbolic relationships.

In theory (assuming very high levels of accuracy are not required), camera calibration is a solved problem [65]. If the correspondence between sufficient points in the image and real world is known, then it is possible to estimate the position and orientation of the camera as well as the lens focal length and CCD aspect ratio. In practice, however, the interpretation of the image to determine corresponding points is not trivial. This is particularly true in the case of cluttered and poorly illuminated real-world scenes. While it is possible to improve the situation by using a hardware solution (building objects that are amenable to machine interpretation) we desire a simple system that does not require any addition to the remote vehicle.

In the teleprogramming system we avoid the difficulty of programming autonomous responses to errors by maintaining the influence of a human operator within the system. Similarly, for camera calibration we avoid the correspondence problem by having the operator manually indicate corresponding features. This is similar to the interactive approach adopted by Kim and Stark [43].

The remote manipulator is calibrated and used as a known reference object. It is moved, under teleprogramming operator control, to a number of positions within the viewing volume of each camera. At each position the operator requests a new high-resolution image from the remote site cameras. As the slave records each image, it simultaneously records the current manipulator joint positions and encodes that information within each image for transmission back to the operator station. This information allows the master station to display each image alongside a simulation of how the remote manipulator should have appeared at that instant in time. The operator may then step through the received imagery, manually indicating corresponding features between the real and simulated views. Through this process, he or she builds a field of points for which both the 2-D position within the image and 3-D position within the model are known. These lists of corresponding points are then used in an implementation of Tsai's algorithm [45] to compute calibration parameters (see Figure 8.1).

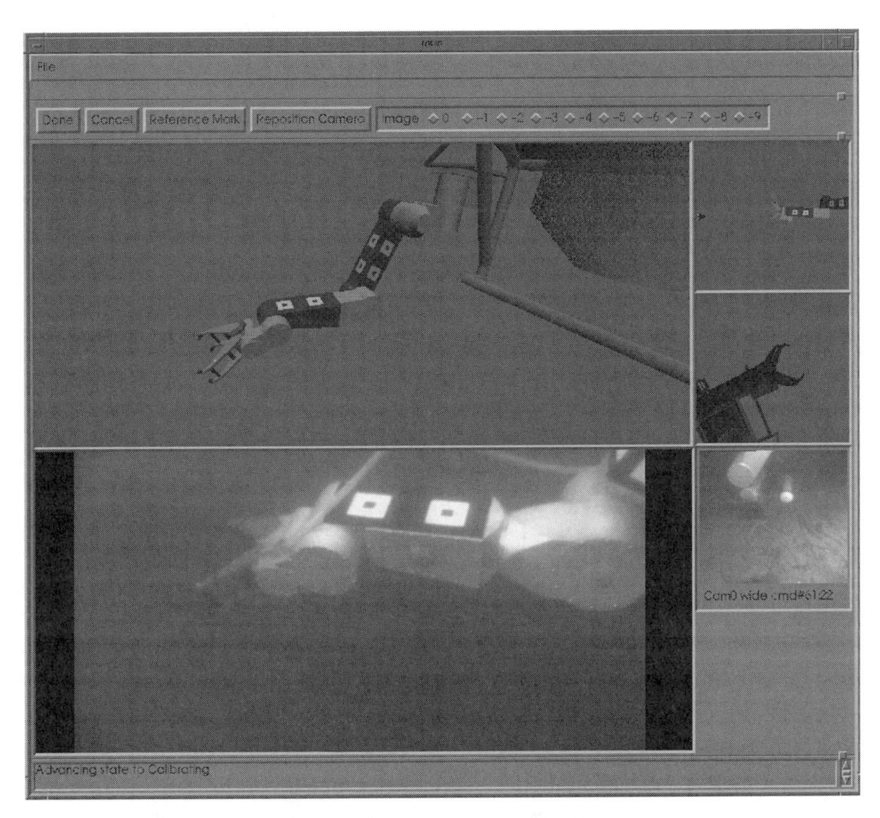

FIGURE 8.1: The interface for camera calibration used during teleprogramming. The system displays for the operator both a received image from the camera to be calibrated (the large lower window), and a simulation of the remote site showing the remote arm at the joint space position it had when that image was taken (the large upper window). By manually indicating corresponding features between several real and simulated views, the operator is able to provide the system with enough information to compute camera calibration parameters. The three smaller windows down the right side show the simulation from two other viewpoints as well as a real image fragment from the remote site.

8.2 Updating the world model

Pictures from the calibrated remote cameras may be overlayed on top of the graphical simulation presented to the operator. This is achieved by using the red dimension of the color space to represent the real images and the blue and green components to show the simulation. The system first sets the location and focal length of the virtual camera to match the calibration data from the real camera. It then projects the corresponding location of the real camera's CCD onto the image plane and performs a linear warping of the real camera image into that space.

The operator may then update the model by simply dragging and resizing the simulated objects until they match the imaged real objects. In many cases the size and shape of the viewed objects are known *a priori*. Thus, the operator need only refine the position of objects in the simulated world (see Figure 8.2).

8.3 Real-time visual imagery

The provision of real-time visual images from the remote site provides a means to reassure the operator that the task is progressing as planned. It also assists the operator in diagnosing, and recovering from, error conditions. Due to the limitations of our communications link, these images must be of comparatively low quality and will not be received at the operator station until several seconds after the time when they were taken. They will therefore serve to supplement, but not replace, the graphical simulation of the remote site.

The bandwidth restrictions imposed by the communications link mean that it is infeasible to transmit every pixel from every image. Thus, we must adopt a more efficient approach.

One possibility is to allow the operator to trade off between frame-rate, resolution, and greyscale [77]; however, the lag introduced by the communications delay makes this undesirable. Also, it is preferable to adopt an automated approach so that the system may continuously react to changes in available bandwidth and current slave actions.

Another option would be to make use of the temporal redundancy between images as in, for example, MPEG compression [62]. However, for the teleprogramming system we desire a more powerful approach. The sys-

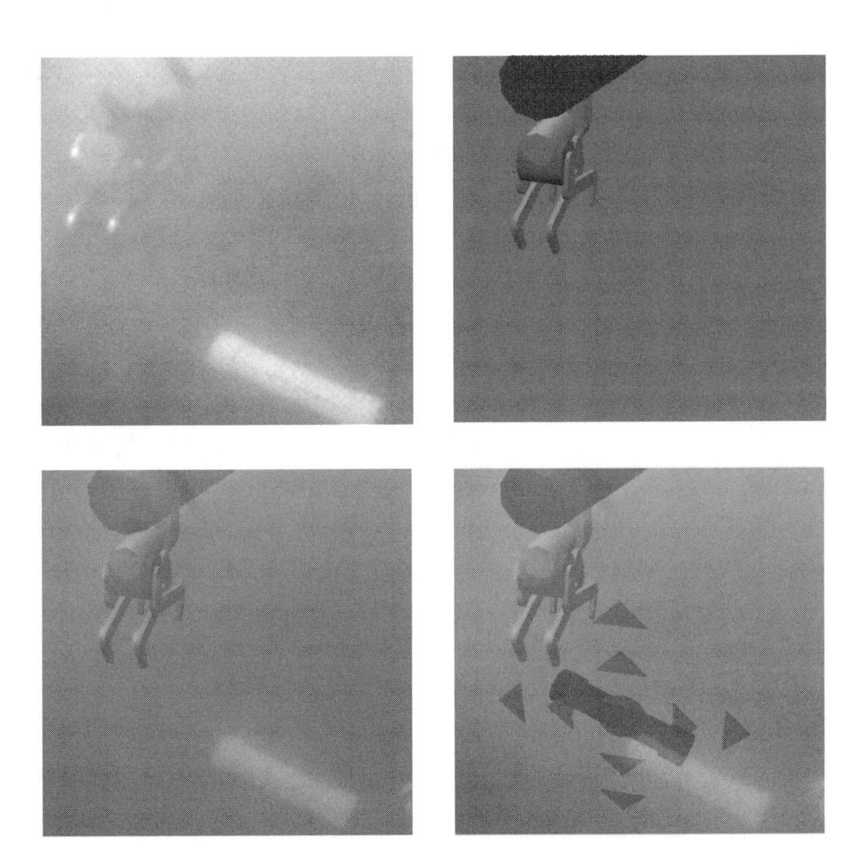

FIGURE 8.2: Once remote cameras have been calibrated, the system may overlay the corresponding real imagery (top left) on the simulation (top right). Any discrepancy should then be visible (bottom left), and the operator may then drag the simulated objects until their position matches that of the corresponding real imagery (bottom right). Here only a single view is shown, but in practice the operator is shown overlayed views from two cameras, so full three-dimensional positions are visible.

tem should react dynamically to changing bandwidth and computational resources and only transmit those pixels that are actually needed.

The following techniques are employed:

- *intelligent fragmentation*—transmit only that fragment of the image that is needed—automatically trade off between sending a large image fragment at a low resolution and a small image fragment at a high resolution.
- *intelligent frame rate*—vary the frame rate dynamically depending on what action is being performed at the slave site – automatically trade off between sending too many images when little is varying and too few images when things are changing rapidly.
- *intelligent task rate*—slow down the slave robot when it moves at a rate in excess of that which the video system can support, given available bandwidth and computational resources.
- *brute force compression*—apply conventional image compression algorithms to the transmitted image fragments.

Traditional bandwidth reduction techniques for image transmission typically rely on pixel-level operations. What we propose are techniques that look not at the image but instead at the scene.

8.4 Intelligent fragmentation

The idea is to select and transmit only that fragment of the image that is "useful" to the operator. This is similar to the sensor planning problem [86]. That is, given a model of the environment, including the robot arm, compute optimal viewpoints for each sensor so as to maximize a prespecified information metric. Abrams et al. have considered the case of offline sensor planning for dynamic environments [1], while Wakita et al. have implemented a simpler online system for land-based teleoperation [92], and McKee et al. have developed a visibility model for the peg-in-hole task [53]. For our implementation the constraints are different. At least in the current system the remote site cameras are fixed. Rather than panning and zooming the remote cameras we are windowing and subsampling their images. The idea is to not only show the operator what is needed, but also to minimize the required communication bandwidth.

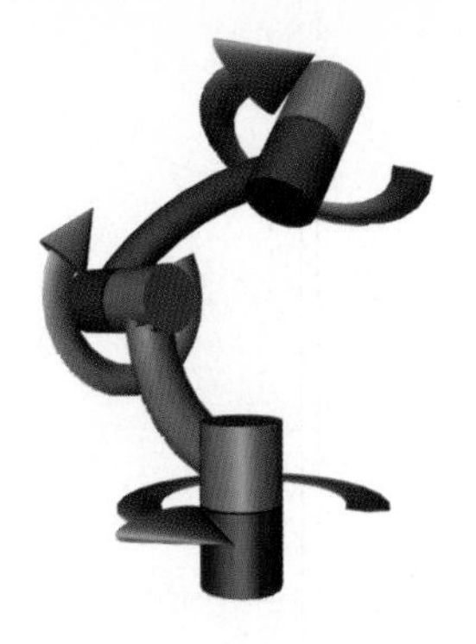

Plate I: Stylized view of a three-DOF robot wrist as used in the Puma 560 factory robot. This style of wrist design is particularly common among commercial manipulators. Inverse kinematics is simplified by having the axes of rotation for all three joints intersect at a single point (see Chapter 2).

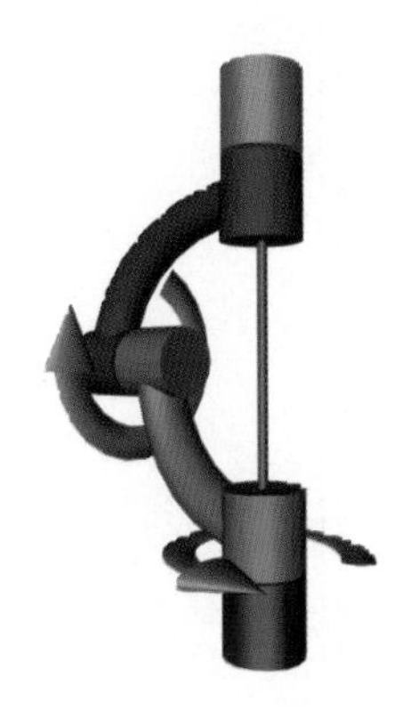

Plate II: Stylized view of a three-DOF robot wrist showing the singular configuration when the first and last joints are aligned. In this configuration only two independent rotations are possible (see Chapter 2).

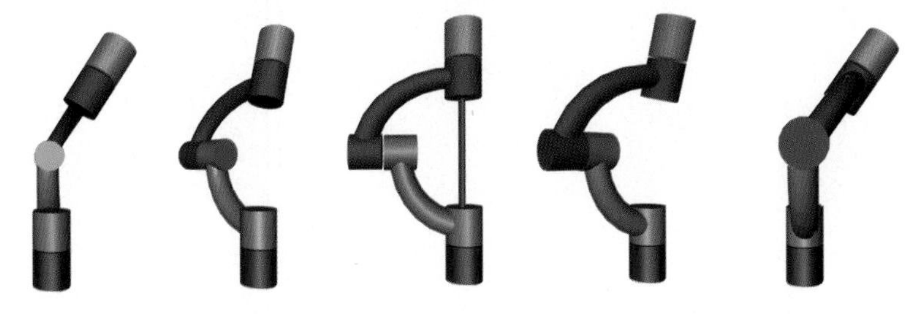

Plate III: Animation of motion between two equivalent Cartesian positions showing redundancy. Two different joint configurations correspond to identical Cartesian orientations, and motion between these two alternative configurations passes through a singularity (see Chapter 2).

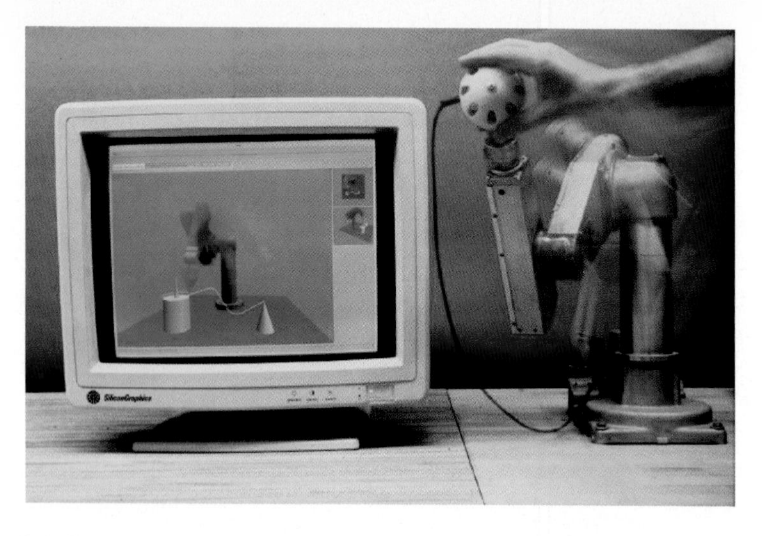

Plate IV: The teleprogramming interface provides the operator with both visual and kinesthetic feedback.

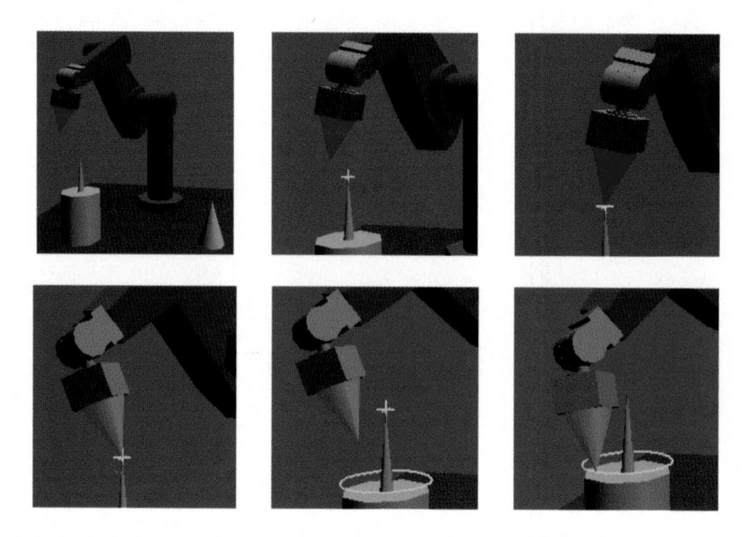

Plate V: In this example, the system aids the operator using two simple synthetic fixtures (see Chapter 7). As the operator moves the end-effector toward the thin cone, the system activates a point-point fixture, guiding him or her toward a perfect point-point contact. The operator then chooses to rotate the end-effector before pulling away and utilizing a point-circle fixture.

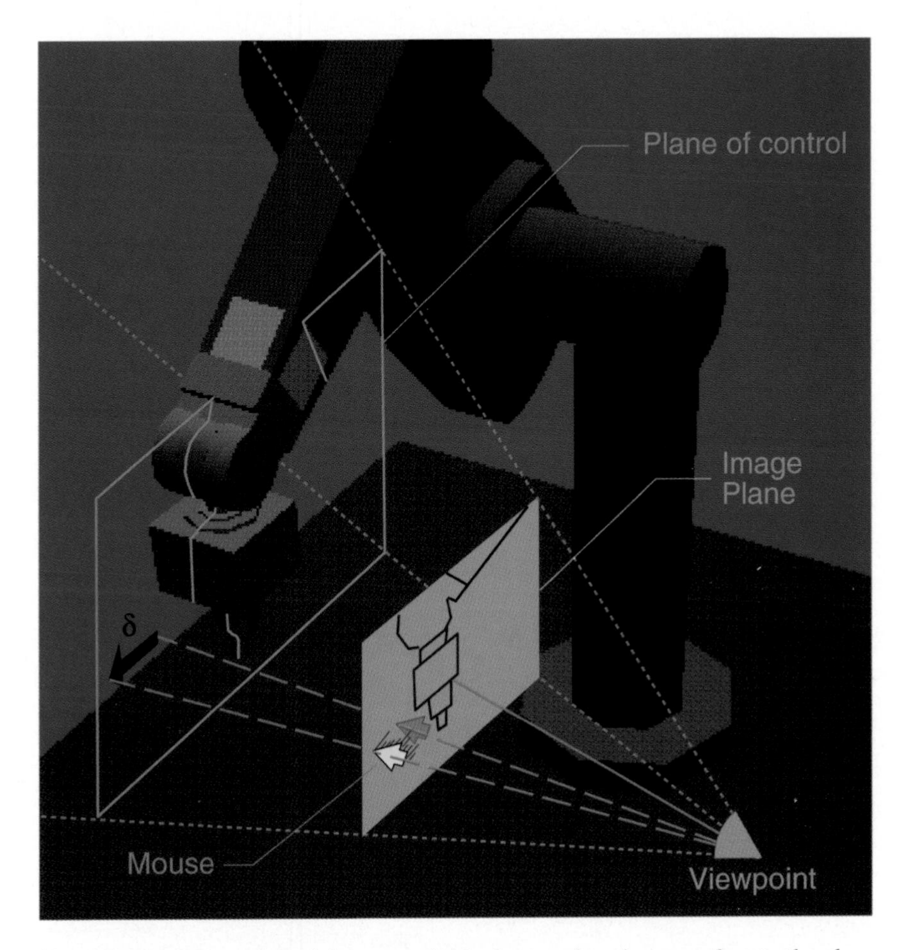

Plate VI: The 2-D motion of the mouse in the image plane is mapped on to the plane of control in the virtual world (see Chapter 6).

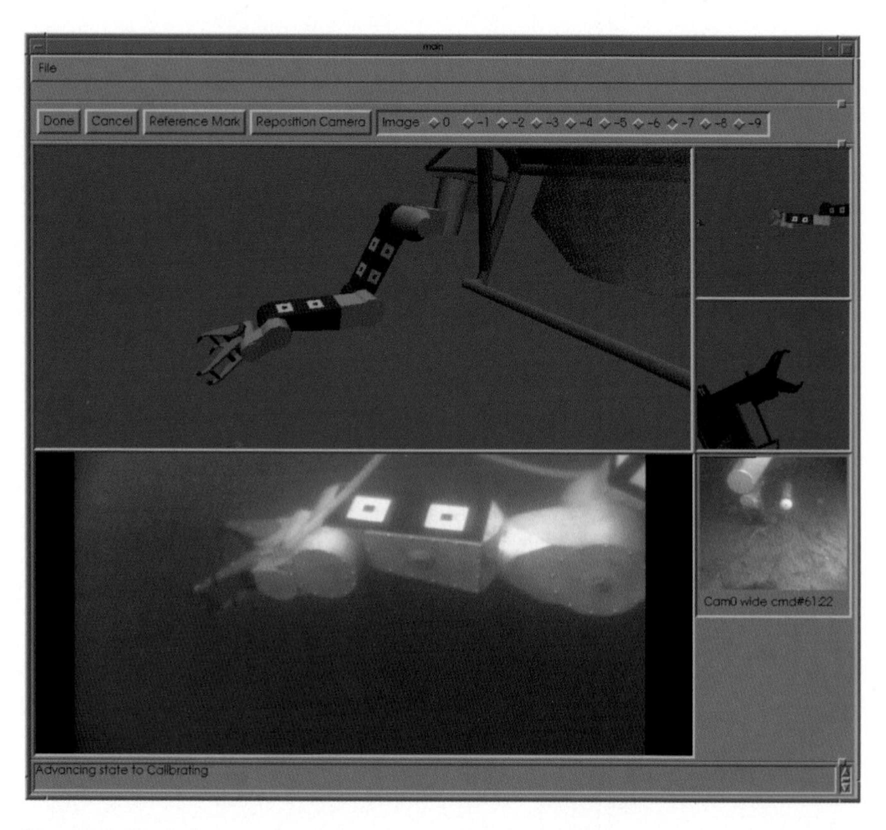

Plate VII: Typical example of the operator interface during camera calibration. The upper windows show simulated views, while the lower windows show actual imagery from the real slave site. The operator will assist the system in performing camera calibration by manually indicating corresponding features between the real and simulated imagery (see Chapter 7).

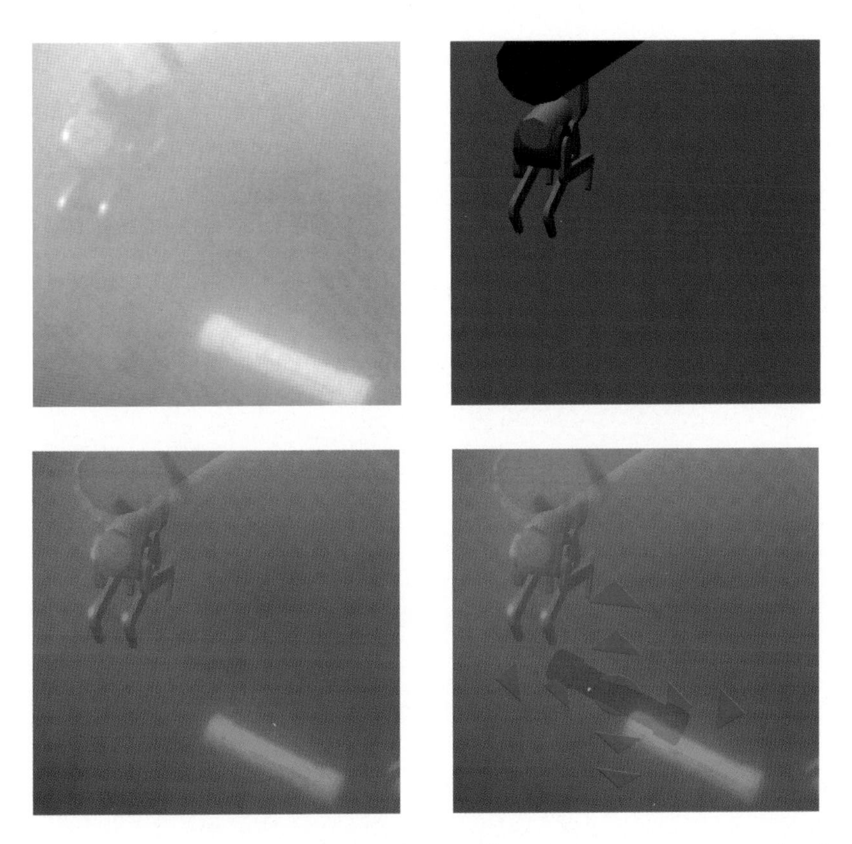

Plate VIII: Once remote cameras have been calibrated, the system may overlay the corresponding real imagery (top left) on the simulation (top right). Any discrepancy should then be visible (bottom left), and the operator may then drag the simulated objects until their position matches that of the corresponding real imagery (bottom right). Here only a single view is shown, but in practice the operator is shown overlayed views from two cameras, so full three-dimensional positions are visible (see Chapter 8).

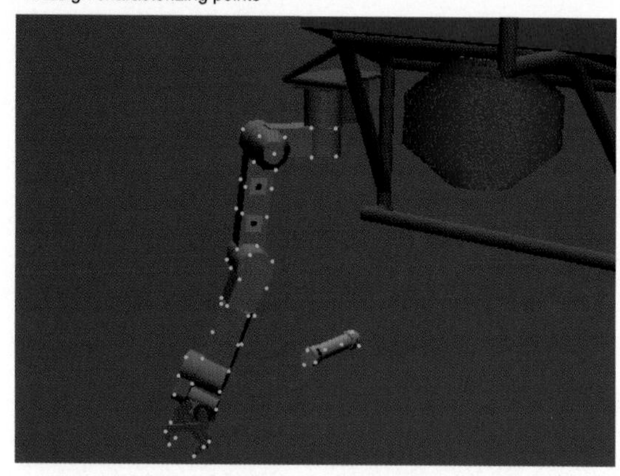

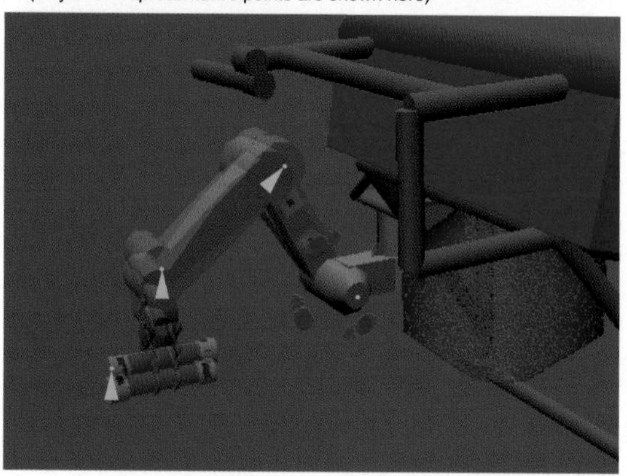

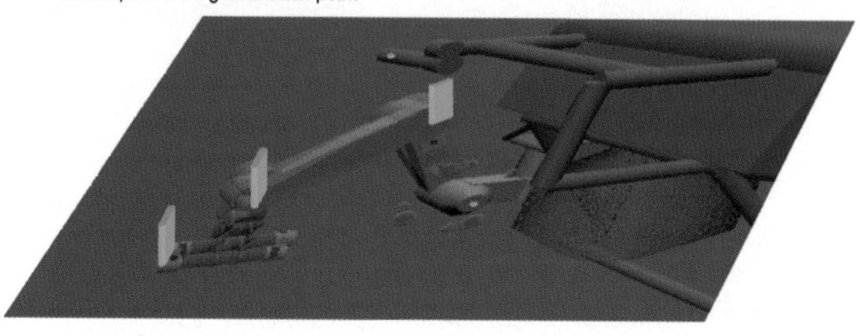

Plate IX: The first step in intelligent fragmentation is to project the operator's commanded action into a model of the remote environment (see Chapter 8).

1.Project the characterizing points into the image plane of each camera

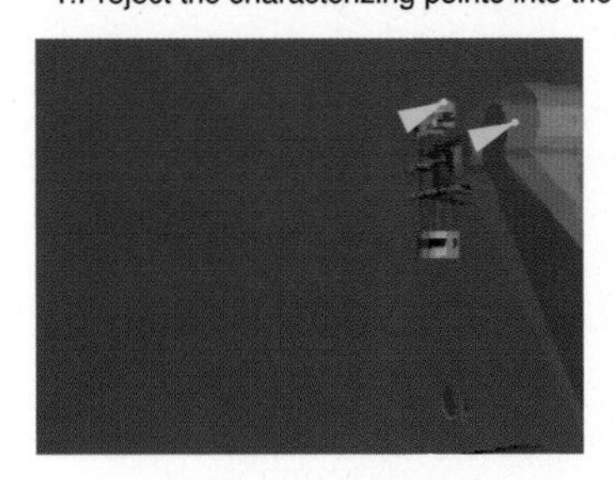

2.Compute a utility metric for each imaged point

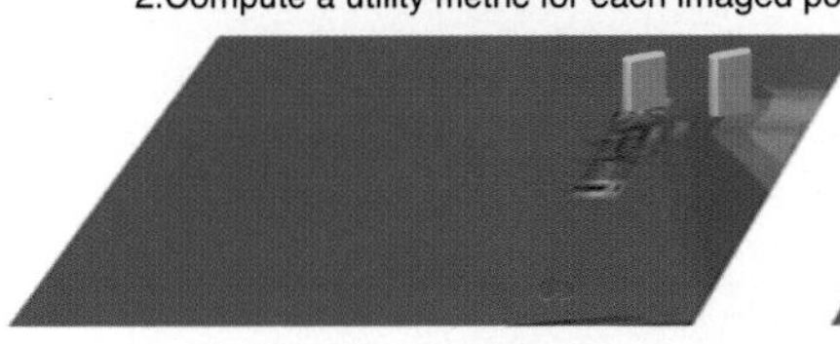

3. Compute the best image fragment for each camera

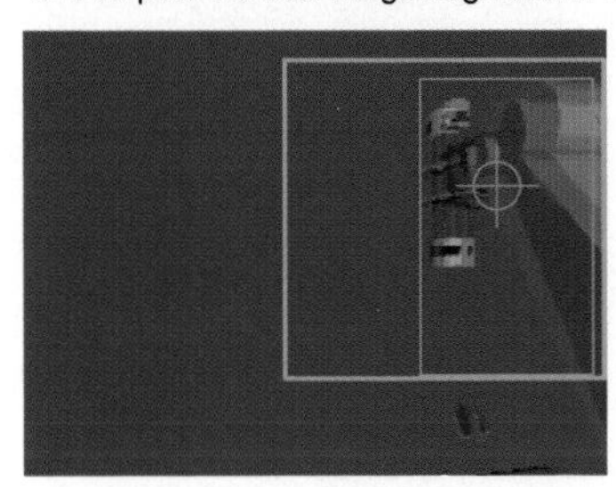

4. Now find the best camera, encode the result within a command and use it to select a single fragment at the remote site

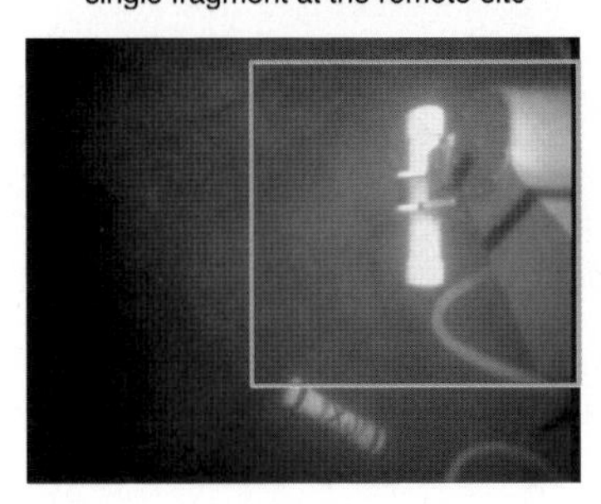

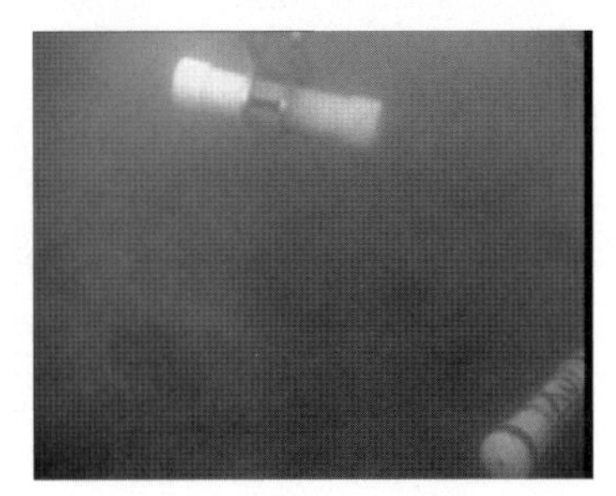

Plate X: The second step in intelligent fragmentation is to project the 3-D motion into the image plane of each camera and compute the single best image fragment (see Chapter 8).

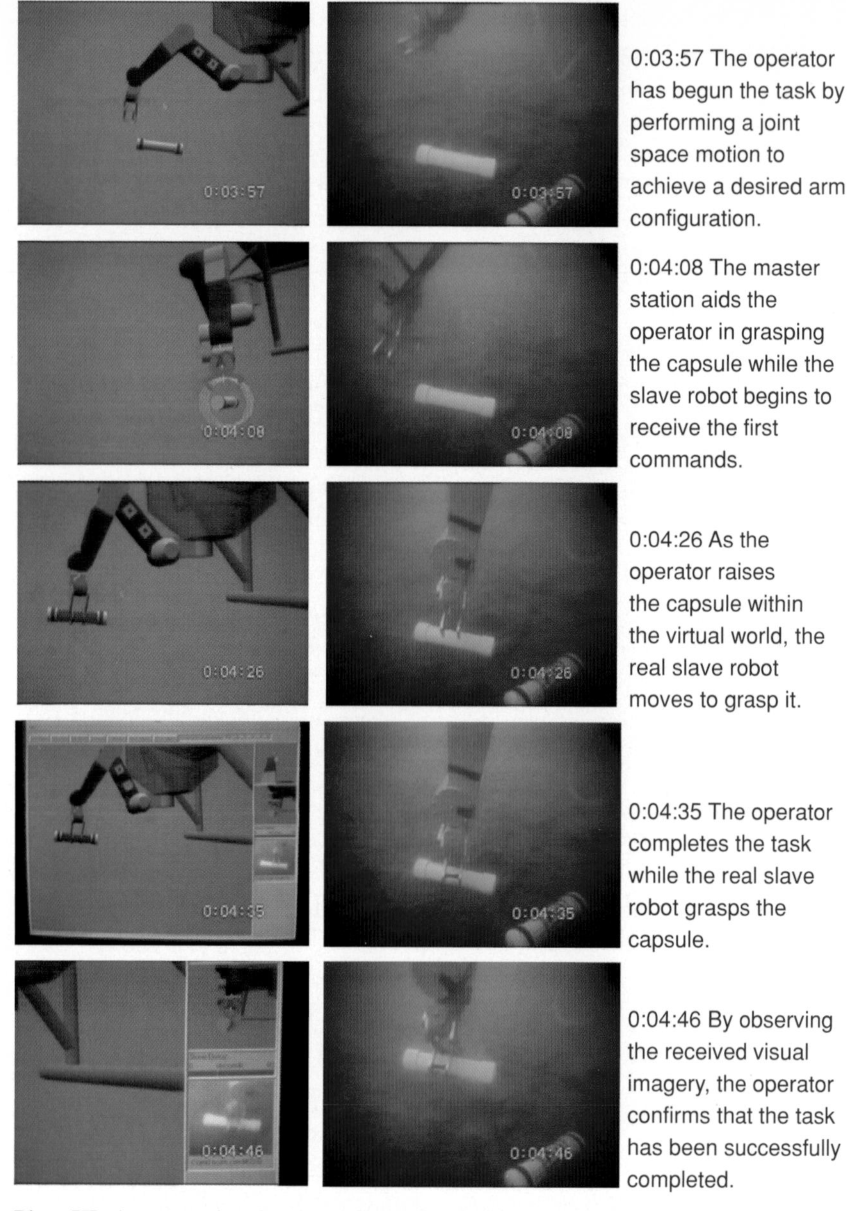

Plate XI: An example of task performance with a 15 second round-trip time delay from the November experiments (see Chapter 11). Stated times are in hours:minutes: seconds.

Finding the "optimal" image fragment requires finding the best fragment within the imaging plane of each camera and then choosing between those cameras to find the single best image fragment.

8.4.1 Finding the best fragment

The operator station predicts, based on the operator's commanded actions, a *region of interest* within the visual field of each remote site camera. The intention is to select that region which would most aid the operator if an error were to occur.

The first step is to project the commanded action into the model of the remote environment and find a 3-D representation for the desired command (see Figure 8.3).

- Define sets of characterizing points within the world model to describe those features that it would be important for the operator to see if an error were to occur. These points represent the "essence" of the model.

  In the current implementation, we define two sets of points. The first, static set, S^s, represents those points which are important even when the manipulator is stationary. This set contains points within the end-effector and any objects with which it may interact. The second, motion set, S^m, represents those points which are important only while they are in motion. This contains points along each link of the manipulator.

$$S^s = \{c_1^s, c_2^s \dots c_{N_s}^s\} \tag{8.1}$$
$$S^m = \{c_1^m, c_2^m \dots c_{N_s}^m\} \tag{8.2}$$

- Through a standard application of transformation math, determine the expected location of each of those points both before and after the currently commanded action. Let $\mathbf{p}(c)$ and $\mathbf{p}'(c)$ be the before and after positions of the characterizing point c.

- Determine a weight for each characterizing point based on its three-dimensional position and/or motion during this command.

  For points in S^s, let $d_s(c)$ be the distance of the characterizing point, c, from the end-effector at the completion of the commanded action,

1.Assign characterizing points

2.Model the motion of those points for the current command
(only a few representative points are shown here)

3.Compute a weight for each point

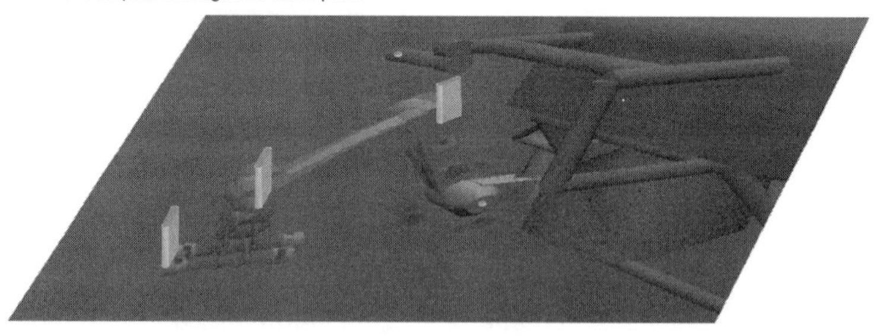

FIGURE 8.3: The first step in intelligent fragmentation is to project the operator's commanded action into a model of the remote environment.

then:

$$d_s(c) = |\mathbf{p}(c) - \mathbf{p}_{\text{effector}}| \tag{8.3}$$

$$W^s(c) = \begin{cases} \frac{T^s - d_s(c)}{T^s} & \text{, if } d_s(c) < T^s\text{mm} \\ 0 & \text{, otherwise} \end{cases} \tag{8.4}$$

For points in S^m, let $d_p(c)$ be the distance by the point c moves, then:

$$d_p(c) = |\mathbf{p}(c) - \mathbf{p}'(c)| \tag{8.5}$$

$$W^m(c) = \begin{cases} 1 & \text{, if } d_p(c) > T^m\text{mm} \\ \left(\frac{d_p(c)}{T^m}\right)^2 & \text{, otherwise} \end{cases} \tag{8.6}$$

The second step is to combine this model of the commanded action with a model of each camera (see Figure 8.4) to compute "optimal" image fragments:

- Using a standard perspective transform, project each of the characterizing points in the 3-D model into the 2-D imaging plane of each camera and determine the potential visibility (ignoring occlusions) of each point. Let $\mathbf{s}(c)$ and $\mathbf{s}'(c)$ be the before and after on-screen locations of the characterizing point c.

- Create a simulated image of the remote site as seen from a camera at the calibrated position and with the calibrated focal length of the real camera. This view is created using z-buffering (so objects in front obscure those behind) and front and back clipping planes (so objects that are too near or too far away are not visible).[1]

  For each point that may potentially be visible, examine the pixel value at the corresponding location within the simulated image to determine if that point is indeed visible. Let V and V' be the sets of all characterizing points, c, where $\mathbf{s}(c)$ or $\mathbf{s}'(c)$ are both within the field of view and not occluded.

- For each point, evaluate a utility function, $U(c)$.

---

[1] The z-buffering handles occlusions perfectly, while the front and back clipping planes provide a crude way to simulate depth-of-field. This is only an approximation since the actual depth-of-field depends on the aperture that the real camera will use—the operator station has no way to know this since it will not be determined until several seconds later when the slave takes the corresponding real picture.

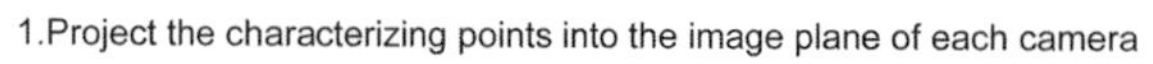

2.Compute a utility metric for each imaged point

3. Compute the best image fragment for each camera

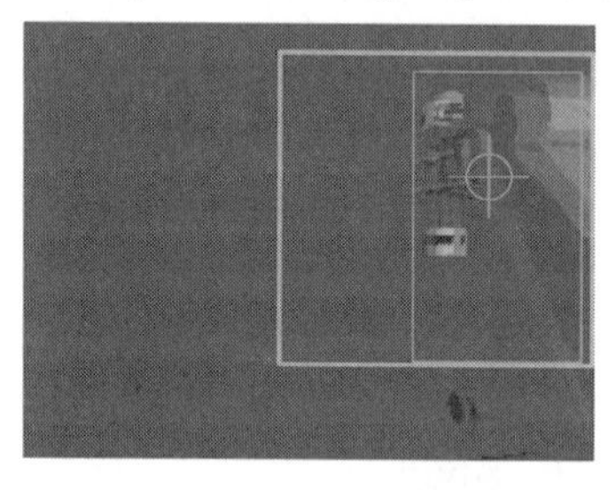

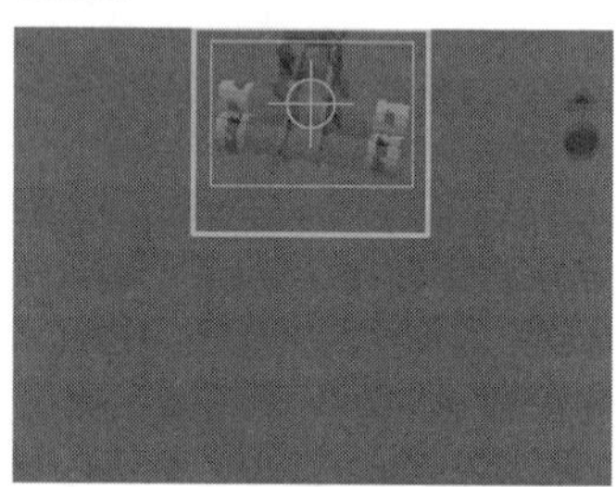

4. Now find the best camera, encode the result within a command and use it to select a single fragment at the remote site

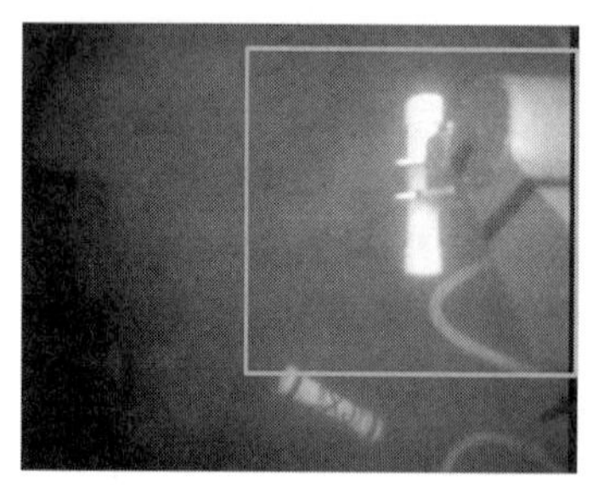

FIGURE 8.4: The second step in intelligent fragmentation is to project the 3-D motion into the image plane of each camera and compute the single best image fragment.

For static points, a point has utility only if it is visible:

$$U^s(c) = \begin{cases} W^s(c) & \text{, if } \mathbf{s}'(c) \in V' \\ 0 & \text{, otherwise} \end{cases} \tag{8.7}$$

For motion points, the utility of a point depends both on its visibility and on how far it is perceived to move upon the image plane. Let d_s be the distance that the point moves upon the camera image plane, then:

$$d_s(c) = \begin{cases} |\mathbf{s}(c) - \mathbf{s}'(c)| & \text{, if } \mathbf{s}(c) \in V \text{ and } \mathbf{s}'(c) \in V' \\ 0 & \text{, otherwise} \end{cases} \tag{8.8}$$

$$U^m(c) = \frac{d_s}{\sqrt{N_x^2 + N_y^2}} \cdot W^m(c) \tag{8.9}$$

- Now compute a bounding box that contains every point whose weight exceeds a predefined threshold and simultaneously compute the "center of mass" of those points. These will be used to compute an "optimal" image fragment.

$$x_{min} = \min(s_x(c) \forall c \text{ s.t. } W(c) > T^w) \tag{8.10}$$

$$y_{min} = \min(s_y(c) \forall c \text{ s.t. } W(c) > T^w) \tag{8.11}$$

$$x_{max} = \max(s_x(c) \forall c \text{ s.t. } W(c) > T^w) \tag{8.12}$$

$$y_{max} = \max(s_y(c) \forall c \text{ s.t. } W(c) > T^w) \tag{8.13}$$

$$m_x = \frac{\sum (U(c) \cdot s_x(c)) + \sum (U(c) \cdot s'_x(c))}{2 \sum U(c)} \tag{8.14}$$

$$m_y = \frac{\sum (U(c) \cdot s_y(c)) + \sum (U^S(c) \cdot s'_y(c))}{2 \sum U(c)} \tag{8.15}$$

Desired window size is:

$$d_x = x_{max} - x_{min} + 1 \tag{8.16}$$

$$d_y = y_{max} - y_{min} + 1 \tag{8.17}$$

Desired center of window is:

$$c_x = \frac{x_{max} + x_{min}}{2} \tag{8.18}$$

$$c_y = \frac{y_{max} + y_{min}}{2} \tag{8.19}$$

- Not every window size may be supported (this is particularly true in the current implementation where only integer subsampling values are implemented). Thus, given this desired window, the idea is to find a practical window size that includes as much as possible of the desired image fragment.

  Let n_x, n_y be the predefined size of the transmitted image fragment and let r_{min} and r_{max} be the minimum and maximum allowable subsampling ratios.

  Now, compute r, the subsampling ratio. If fractional r values are acceptable, then:

$$r = \min(\max(\frac{d_x}{n_x}, \frac{d_y}{n_y}, r_{min}), r_{max}) \quad (8.20)$$

  otherwise:

$$r = \min(\max(\text{ceil}\left(\frac{d_x}{n_x}\right), \text{ceil}\left(\frac{d_y}{n_y}\right), r_{min}), r_{max}) \quad (8.21)$$

  Thus, we now have a practical window of size $(r \cdot n_x, r \cdot n_y)$.

- Now, make use of any difference between practical and desired window sizes to bring the center of the practical window closer to the center of mass of the image and perturb the image center (if necessary) so that the chosen window does not exceed the imaged frame.

  A practical value for the center of the imaging window is:

$$c_x = \begin{cases} c_x & \text{, if } r \cdot n_x = d_x \\ c_x + \min(\frac{d_x - r \cdot n_x}{2}, |m_x - c_x|) \cdot \\ \text{sign}(m_x - c_x) & \text{, if } r \cdot n_x < d_x \\ \min(N_x - \frac{r \cdot n_x}{2}, \max(\frac{r \cdot n_x}{2}, c_x + \\ \max(r \cdot n_x - d_x, |m_x - c_x|) \cdot & \text{, if } r \cdot n_x > d_x \\ \text{sign}(m_x - c_x))) \end{cases} \quad (8.22)$$

$$c_y = \begin{cases} c_y & \text{, if } r \cdot n_y = d_y \\ c_y + \min(\frac{d_y - r \cdot n_y}{2}, |m_y - c_y|) \cdot \\ \text{sign}(m_y - c_y) & \text{, if } r \cdot n_y < d_y \\ \min(N_y - \frac{r \cdot n_y}{2}, \max(\frac{r \cdot n_y}{2}, c_y + \\ \max(r \cdot n_y - d_y, |m_y - c_y|) \cdot & \text{, if } r \cdot n_y > d_y \\ \text{sign}(m_y - c_y))) \end{cases} \quad (8.23)$$

8.4.2 *Finding the best camera*

Having determined the best image fragment available from each camera, we must now choose between those fragments to find the single best one. To make this decision, we consider visibility, operator preference, and continuity.

Visibility metric

A metric for the utility of the chosen image fragment may be evaluated by considering the proportion of the total action "weight" that is visible.

$$M_s = \frac{\sum U^s(c)}{\sum W^s(c)} \tag{8.24}$$

$$M_m = \frac{\sum U^m(c)}{\sum W^m(c)} \tag{8.25}$$

Operator metric

To permit the operator some control over the chosen camera view, we provide him or her with the ability to bias the system toward a particular camera by manually selecting a metric, M_o. A value of greater than zero biases the system toward a particular camera, while a value less than zero has the reverse effect.

Complete metric

Combining visibility and operator preferences gives:

$$M = C_s \cdot M_s + C_m \cdot M_m + M_o \tag{8.26}$$

Where C_s and C_m are constants used to adjust the relative weight of each metric.

The best fragment is now that which maximizes this metric.

Continuity considerations

The preceding discussion emphasized quantitative measures for selecting images. However, the best images are not necessarily those with the highest quantitative score. In this section, consideration is given to the more subjective and qualitative aspects of providing imagery for the human operator.

The provision of "live" imagery during task execution is analogous to the selection of different shots for a live television program. This process is known as switching, or instantaneous editing [99]. The goal of such editing is not only to provide the viewer with maximal information, but also to do so in a way that allows the viewer to "follow the action." Thus, choosing the best camera to use next depends not only on what each camera sees, but also on which camera's view is currently being shown. The idea is to switch between views while maintaining both position and motion continuity. Positional continuity refers to objects being seen in the same relative positions, while motion continuity refers to objects in motion appearing to move in the same direction from scene to scene.

For example, if two cameras were placed on opposite sides of the robot and it moved along a path between the two cameras, then switching camera views could disorient the operator by causing the motion to suddenly switch directions (see Figure 8.5).

The difference in perceived motion direction and object location depends largely on the relative orientation of the cameras. For film, the rule for camera positioning is relatively simple. An imaginary line is inscribed along the direction of motion, then all shots must be made from cameras on one side of that line. This is the "180 degree rule" [54, 61]. For our application, consideration must be given to cameras with different heights as well as orientations. Furthermore, we must still send imagery to the operator even when no good view is available. Thus, a more general metric is required:

$$M_c = \frac{2 + \mathbf{n}_z^c.\mathbf{n}_z^p + \mathbf{n}_x^c.\mathbf{n}_x^p}{4} \tag{8.27}$$

where $\mathbf{n}_z$ and $\mathbf{n}_x$ are normals describing the viewing direction and orientation (in world coordinates) for both the current (c) and prospective (p) cameras.

Another form of continuity is consistency between the operator's chosen view of the simulated remote site and the real image fragments returned from the real remote site. In cases where camera views have similar utility, there may be an advantage in choosing that view whose viewing direction most closely matches the operator's chosen viewpoint within the virtual environment.

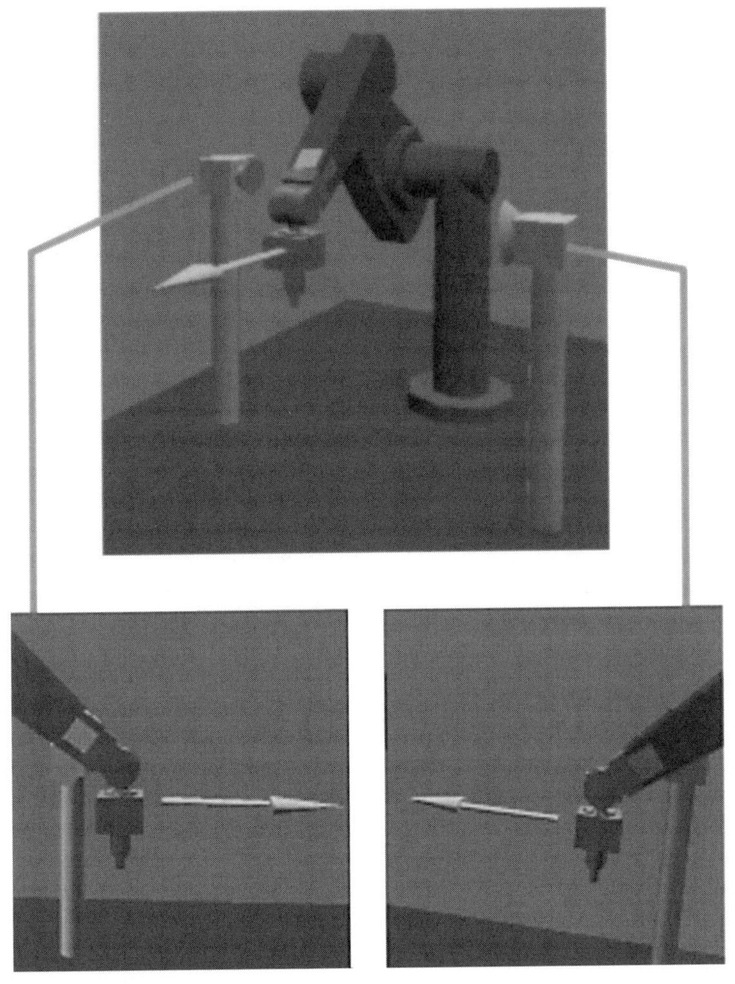

FIGURE 8.5: Switching between different camera views can break motion continuity, causing disorientation. In this case motion switches from left-to-right on the first camera to right-to-left on the second.

In the current implementation, where there are only two cameras, there is so little choice that visibility is the dominating factor and the continuity metric is not used. However, one could imagine a situation where dozens of inexpensive cameras were fitted to the vehicle and, in that case, relatively minor effects such as continuity will play an important role in disambiguating between cameras with similar visibility metrics.

8.4.3 Encoding the desired image fragment

The desired image fragments and utility metrics are encoded within each generated command. During task performance, the slave compares the metric for each operational camera, finds the one with the largest score and selects the desired image fragment from that camera for compression and transmission to the operator.

In the current system, there are two remote site cameras, each with a resolution of 720 x 480 pixels. The real-time images, created by windowing and subsampling based on the *region of interest*, are 192 x 160 pixels in size (see Figure 8.6).

8.5 Intelligent frame rate

In most real-time vision systems the frame rate is constant. However, this need not always be the case. Particularly when images are being transmitted over a limited bandwidth channel, there is an advantage in sending only those images that are actually necessary.

The frame rate can be varied based on the rate of motion of the slave manipulator. This makes it possible to send fewer images while the system is changing slowly, and more images when rapid events are in progress. In the present implementation, the slave bases its update rate on the frequency with which commands are executed. This works well for Cartesian motions since, in that case, the frequency of the generated commands is directly related to the operator's desired rate of motion. However, it is not optimal for joint space motions where a single command could result in a large motion at the slave.

There are two solutions to this. The first is to simply not permit the operator station to send large joint space motions — instead forcing it to break them up into a sequence of smaller actions. This is simple to implement

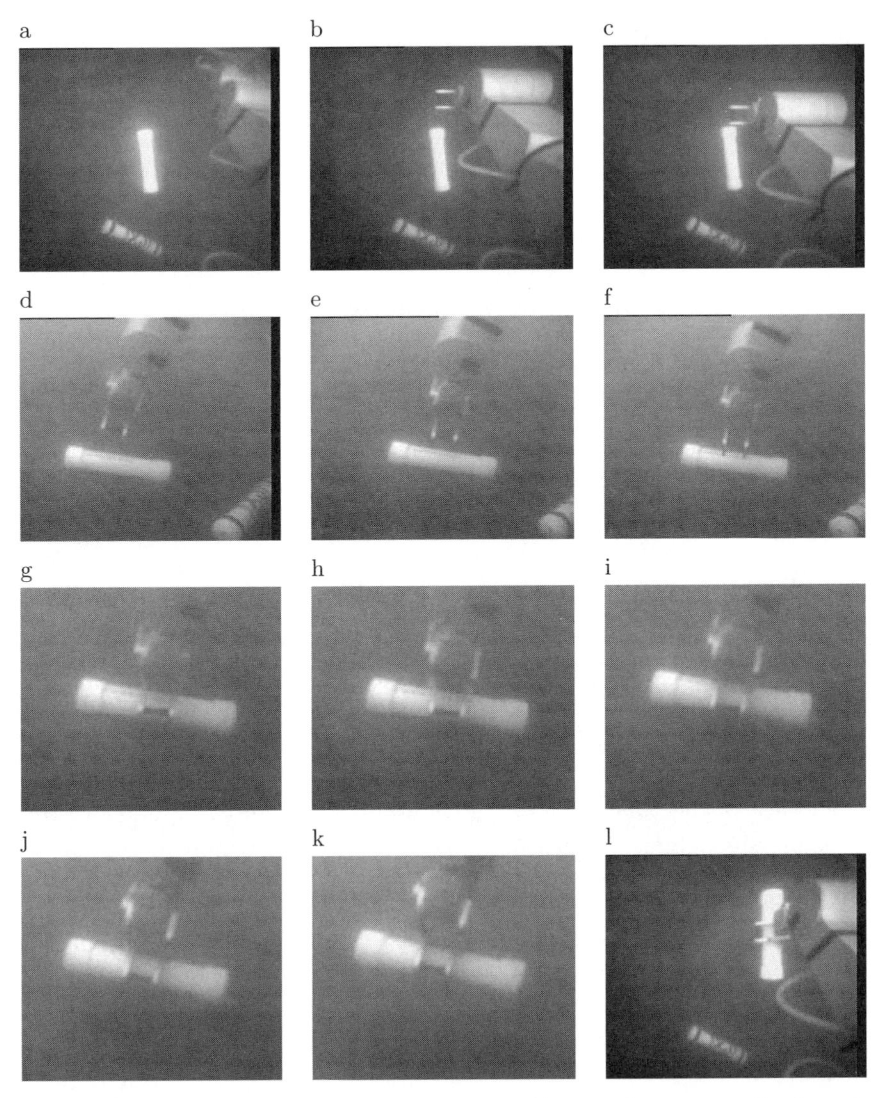

FIGURE 8.6: Examples from a sequence of image fragments displayed to the operator during subsea experiments conducted off the coast of Massachusetts in November, 1994. Note how the view automatically switches cameras, pans, and zooms to follow the action at the remote site. These images were chosen using an earlier implementation of the visibility algorithm that did not consider occlusions.

but it wastes bandwidth. Another, more powerful, alternative is to encode a desired frame rate along with each command. This would have the added advantage of allowing the master station to request high update rates during motions in which errors are likely to occur (such as gripper closure). The frame rate could be encoded as a measure of the desired inter-frame time. More formally:

Definitions:

- t_a be the actual time elapsed since the image grab/compress process last began to grab an image
- t_{d_i} be the desired inter-frame time for the i'th command
- t_{d_m} be the maximum allowable inter-frame time

Assume an image was snapped during command j, then a new image should be snapped during command k, where $k \geq j$, iff:

$$t_a > \min\{t_{d_m}, \min\{t_{d_i} : j \leq i \leq k\}\}$$

8.6 Intelligent task rate

In all current teleoperation implementations, the connection between imaging and manipulation systems is unidirectional. The camera system may react to the robot's motion, but the robot does not react to the cameras. The slave manipulators perform tasks the same way and at the same speed regardless of the bandwidth and computational resources available to the imaging subsystem. We propose that there is good reason to consider a closer bidirectional coupling.

Images of the robots are necessary if the operator is to have adequate information to diagnose execution errors at the remote site. Thus, if the frame rate is too slow, then there is a tradeoff between continuing with the execution anyway (and risking that an error will occur and then be misdiagnosed), and pausing (and wasting time).

The solution adopted in the current system is an exponential backoff algorithm. As the inter-frame time increases, the system inserts increasingly large delays between executed commands. (An improved system would be to slow down each command but the current infrastructure does not readily support this option.)

An implementational detail of the current system is that both the image snap/compress process and the command interpreter are running on the same processor. This means that adding delays into the interpreter loop doesn't just slow down the interpreter—it also has the desirable positive effect of speeding up the imaging process. This allows the system to dynamically trade off limited computational resources between the two competing processes.

8.7 Compression algorithms

The above methods limit the size of the transmitted image and reduce the average rate at which image fragments must be transmitted. A significant additional reduction in bandwidth can be achieved by compressing the transmitted image fragments. Image compression has been investigated by a number of researchers, and several well-known algorithms exist [62]. In the current system, JPEG compression is employed. More recent techniques, such as wavelet compression [35], may offer better performance.

Tests on preliminary images[2] indicate that, while high compression ratios cause noticeable image degradation, a twenty-fold reduction is possible while still leaving sufficient information in the image to enable task performance.

It may be possible to further improve compression performance by making use of temporal redundancy—using the non-motion of much of the image to aid in compression. However, this is problematic for underwater teleoperation applications, since the motion of either the vehicle or particulate matter in the water can make it appear that almost every point in the image is moving. Furthermore, since we are panning, zooming, and switching cameras it will often be the case that one image and the next are very dissimilar.

8.8 Other sensory modalities

In this chapter, the emphasis has been on visual sensors at the remote site. Cameras happen to be convenient because their behavior is well understood and they are relatively inexpensive. However, the techniques described here should be equally applicable to other sensors. For example, if the remote

[2]These tests were performed by David Hoag, Northeastern University, using images supplied by the author.

robot were equipped with a distance sensor, we could predict the utility of its information by projecting its imaging volume into the world model in much the same way as for the cameras.

In cases where a movable sensor is used, a possible variation on the above technique is the following approximation:

- Discretise the sensor motion to find a set of possible locations.
- Imagine there is one virtual sensor at each discrete location.
- Apply the above algorithm to find which of those virtual sensors has the most utility.
- Command the single real sensor to move to the location of that virtual sensor.

8.9 Future implementations

It is now feasible to consider having several dozen cameras at the remote site. Adding additional cameras does not cost additional bandwidth or power (since only one is in use at any instant) and modern cameras are small, lightweight, and relatively inexpensive.

Replacing pan-tilt units with multiple fixed cameras would provide greater reliability, redundancy and, in the case of teleprogramming, the opportunity to obtain multiple images of the same event from several different angles.

Since the system already predicts which camera will provide the best view, it should be quite feasible to extend it to consider lighting as well. Rather than flooding the remote site with light (as is often done now), we could instead employ a number of illumination sources and activate them selectively based on the chosen camera, arm, and expected remote site configuration.

In the current implementation, the determination of which image fragment is best depends only on what the operator station expects should be visible. This has the natural advantage that the slave can choose between available cameras without any need to examine images from those cameras. One interesting possibility would be to have the slave assign additional metrics for each camera based on current visibility and lighting considerations. Then the choice of which camera to use could be biased away from the operator station's prediction to favor cameras with sharper images.

8.10 Summary

In the teleprogramming system, the bandwidth between sites is too low to allow transmission of every piece of visual imagery, while the communications delay and desire to keep the operator immersed in their virtual world makes it inadvisable to require the operator to choose which visual imagery should be transmitted. Thus, intelligent techniques are required. Three were introduced: intelligent fragmentation, intelligent frame rate, and intelligent task rate.

To perform intelligent fragmentation we first define characterizing points to represent those features of the world model that may be of use in diagnosing an error. Then, during task execution, we determine a weight for each characterizing point based on its relevance to the currently commanded action. Next, we find the corresponding locations of those points within the image plane of each camera and compute a utility metric for each based on its weight and visibility. Then for each camera, we compute that image fragment that will be of maximal utility and finally find the single best image fragment from among all possible cameras.

An intelligent frame rate requires varying the rate at which frames are taken to match the rate of task performance, while intelligent task rate is almost the reverse—it requires limiting the maximum rate at which the task may be performed to match the maximum rate at which images can be taken.

In the current implementation, the emphasis has been on quantitative measures (such as visibility). In future implementations it can be expected that the rate with which images may be transmitted will increase (due to better image compression techniques and greater communication bandwidths). It can also be expected that more cameras will be used (due to decreasing cost of imaging hardware). These two factors mean more imagery will be available at higher rates and this, in turn, will place an increasing emphasis on more qualitative measures, such as motion and positional continuity.

9

Expecting the Unexpected

> "The smart thing is to prepare for the unexpected."
>
> —*from a Chinese fortune cookie*

It is tempting, particularly when working in a laboratory, to avoid the problem of errors entirely by just pretending that nothing unexpected will ever happen. This is like crossing the road with your eyes closed. It might work the first few times, but that is no reason for confidence. If our system for controlling a remote robot is to be practical, then we must have an effective means for dealing with the unexpected.

In this chapter, we'll begin by looking at ways to reduce the frequency with which errors occur, then look at ways to mitigate the effects of likely errors and then finally examine techniques for diagnosing and recovering from those errors that occur.

9.1 Definition

For purposes of this discussion, an error is considered to be any situation in which the behavior at the remote site does not match that desired by the operator. Even if the error was caused by the operator performing the wrong action it is still an error. In this text the terms "error event," "unexpected event," or "exceptional event" are all synonymous.

9.2 Avoiding operator error

The first, and perhaps most obvious, technique is to avoid errors by dissuading the operator from performing actions that are likely to fail when attempted at the remote site. In the present implementation, this is achieved through provision of active force and visual clues (see Section 7.7). As the operator interacts with the virtual slave site, the system actively encourages him or her to perform actions that are likely to succeed (such as contacting a surface whose position is well defined), while dissuading him or her from actions that will likely fail (such as sliding along an edge whose location is uncertain).

Dentist: "What seems to be the problem?"
Patient: "It hurts when I press here."
Dentist: "Well, then don't press there!"

In addition, because we maintain a human operator "in the loop," we gain the benefit of his or her experience. Operators will quickly recognize situations that are likely to produce errors and modify their behavior accordingly. It might seem obvious, but the impact of having a human who learns from experience should never be underestimated.

9.3 Avoiding interpretation errors

In the teleprogramming system, the operator is not writing a robot program by hand. Instead he or she is interacting with a virtual slave site and, based on that interaction, the system infers which commands he or she wishes executed by the real remote robot. If the system's interpretation of operator action does not match that expected by the operator, then an error results. These errors are just as costly as something unexpected happening at the remote site but are at least easier to prevent.

Once again, we rely on synthetic fixtures to assist in avoiding errors. By forcing the operator to make deliberate, unambiguous actions, they simplify the task of turning those actions into commands while simultaneously allowing the system to make its expectation and interpretation clear to the operator.

For example, consider the case of sliding the end-effector on a surface and assume there were no force and visual clues. Since the operator's motion of the master arm will never be perfect, the system will continually have to guess whether or not motion just slightly off the surface was intentional. Through the use of synthetic fixtures, much of this ambiguity can be removed. The force clues actively hold the end-effector in contact with the surface, and moving away requires that the operator overcome a distinct resistance. This action is thus very obvious—both to the system and to the operator. It's now clear to the system what command the operator intended, and it's clear to the operator what command the system will generate.

9.4 Predicting errors

Consider now the interesting case where we know, in advance, that an error is likely to occur. Given that knowledge, it should be possible to reduce its severity.

The first method is to direct remote site sensors so as to increase the likelihood that the error is detected as soon as possible after it occurs. Assuming that we can't prevent the error from occurring, it should at least be possible to ensure it is not compounded by attempting to execute subsequent commands. If remote site sensors are inadequate, then they may be supplemented with active techniques. For example, we could perform a motion that would be expected to fail if, and only if, the error occurs. In this way the operator may continue with the task knowing that the slave will only continue if the error does not occur.

An even better solution is to precode a recovery action within the generated command stream. In this case, it may be possible for the slave to detect and (using the preprogrammed solution) recover from the error without the operator even being aware that it has happened. It is for this reason that the teleprogramming language includes "if" and "goto" statements (see the next Chapter for details).

The above techniques will work and are useful in limited situations, but not all errors may be predicted *a priori*. Rather than attempting to program around exceptional conditions it is much better to devote limited resources to building a robust system that enables task completion despite errors.

9.5 Error detection and diagnosis

In the teleprogramming system, we rely primarily on autonomous error detection. This allows the operator to continue working after a command has been generated without needing to wait and observe the results of that command. To be effective, autonomous detection has to be perceived as reliable by the operator. In the teleprogramming system, exceptional conditions are detected at the slave site by comparing actual sensory data with that predicted by the master station (and encoded symbolically within transmitted commands).

Error diagnosis is shared by the system and the operator. The system notifies the operator as to when the error was detected and how it was detected. It also provides some suggestion as to the cause of the error. This is implemented as a look-up table at the slave. For a particular error, while in a particular behavioral state, there is a hard-coded "best guess" as to what may have caused the error. The system does not make any attempt at performing complicated reasoning about what caused the error—that is the operator's job.

When an error is detected at the remote site, it signals the master site. Once that message arrives at the operator station, it "winds back time" to the point where the error occurred. The effect for the operator is very much like watching a movie in reverse—all of the actions performed after the error command was transmitted are undone. This animation is intended to smoothly transition the operator from thinking about the future, to thinking about what he or she was doing several seconds in the past.

The operator is then apprised of the cause of the error by both a text message and, in some cases, a visual clue on the screen. An example text message is "Unexpected force detected in +Z dirn" and in this case, an arrow is overlayed on the graphics indicating the +Z direction for the frame in which the error occurred. The operator has the option, at this point, of stepping either backwards or forwards through the generated commands. For each command, displays are available of the preliminary error diagnosis, the commanded slave positions, the actual recorded slave positions, and the recorded real visual imagery.

Since each transmitted command is only for a single, short (typically around 500 ms) action, then knowing which command generated the error message gives the operator some indication as to causality—there is only so much that can go wrong with each simple command. However, this information, coupled with the slave's preliminary diagnosis, is not usually sufficient for the operator to diagnose and correct the error, the reason being that the slave's responses may, themselves, contain errors.

In general, for each command there are five mutually exclusive possibilities:

- *true success*: No error occurs and none is reported. If we were operating in a perfect world, this is all that should ever happen.
- *true error*: An error occurs and is correctly reported. If one assumes that some errors will inevitably occur, then optimal performance will be realized only if they are immediately detected and correctly diagnosed.
- *false success*: An error occurs but none is reported. In this case, the slave fails to detect an error even though one actually occurred. This is the worst possible situation since, when an error goes undetected, the operator and system will continue with the task. By the time the problem is noticed (usually as a result of some later dependent operation failing), many commands may have been sent and executed and the operator is faced with the difficult task of figuring out which possibilities may have led to the detected error condition.
- *false error*: No error occurs but one is erroneously reported. These waste time and can be problematic since they habituate the operator to ignoring error messages.
- *wrong error*: An error occurs but a different error is reported. This is the most benign of the erroneous responses. The operator is well aware that the system's preliminary diagnosis of the error is only a "best guess."

As a result of these different possibilities, the operator must look not only at what may have caused the reported error, but also at whether that error was itself erroneous. To do this, the operator needs more information.

The comparison of actual and commanded positions just prior to the point where the error was detected is perhaps the least-cost form of error analysis. To efficiently support this, the slave sends positions in two different ways. First, as commands are executed it is continually sending status information back to the operator station. This provides trajectory information that is spatially accurate but relatively sparse temporally (only one position is sent for each executed command). When the slave detects an error, it follows the error signal with a more detailed record of the positions during the previous few seconds of motion.

This positional information is very useful in analyzing frequently occurring errors since the operator becomes accustomed to the characteristic "positional signature" for these errors. For example, in the case of lowering a bolt into a hole, if the commanded position would have put the bolt inside the hole, but the actual position shows only the end of the bolt on the surface, then it is likely that the slave missed the hole and is pushing the end of the bolt against the surrounding surface.

Pictures of the real remote site provide the operator with an excellent tool for diagnosing errors (see Figure 9.1). The master station buffers a number of recent images received from the remote site. These are indexed by command number and, as the operator moves forward and back through the historical record of executed commands, the appropriate visual image is displayed. In some cases the single, most-recent, image is sufficient to characterize the error, while in other cases it is the time history that is more useful.

Full resolution images (rather than the image fragments transmitted during task performance) may be necessary after an error for use in recalibrating the world model. Thus, when an error is detected, the slave immediately takes a full picture from each camera for transmission to the operator station.

9.6 Error recovery

In recovering from an error, the operator has the option of editing the world model to bring it into closer alignment with the real world. He or she also has the option of overriding the system's assumptions about the success or failure of previous commands. The operator can tell the system that any previous command failed (even though the system may have "thought" that it succeeded) or that any previous command succeeded (even though the system may "think" that it failed). This ability is important because the state of the world model influences the system's ability to aid the operator. For example, if the system thought the slave had unscrewed a bolt but the operator determined that to be false, then the operator must be able to go back to that command, correct the system's assumption, and then continue

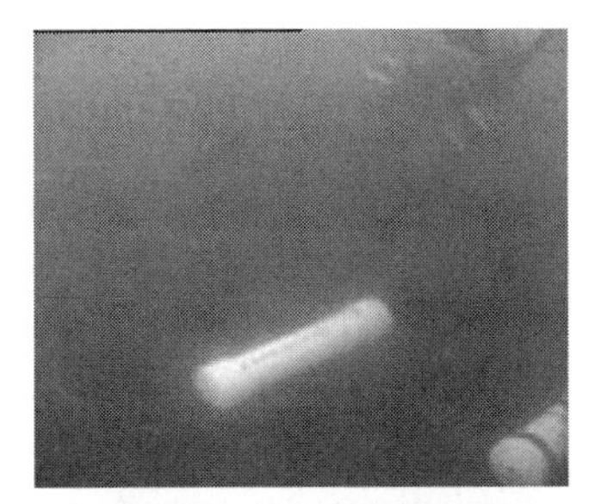

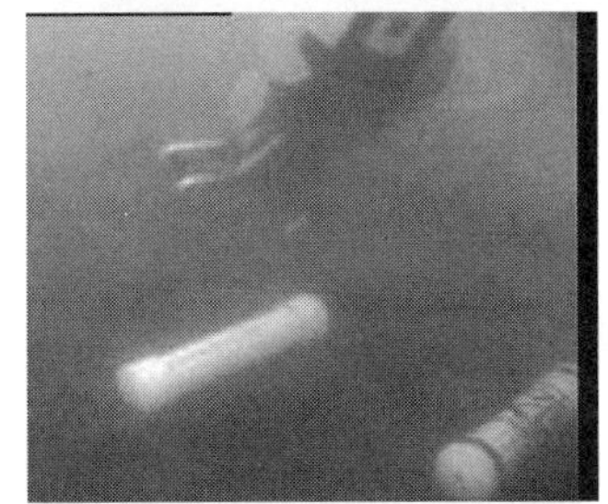

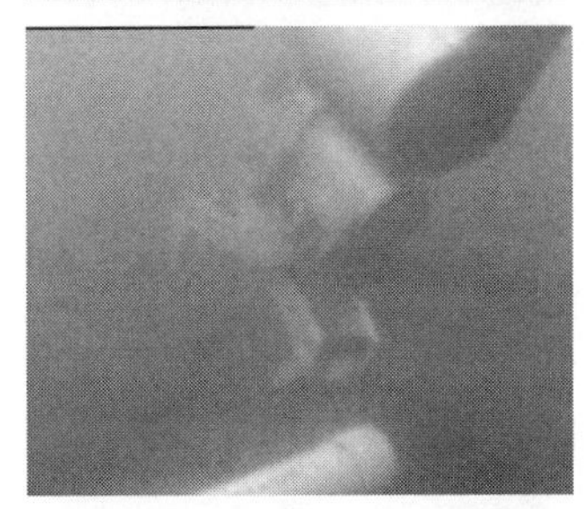

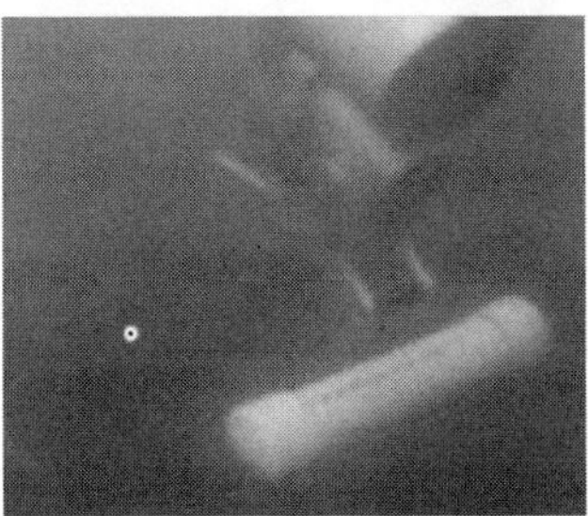

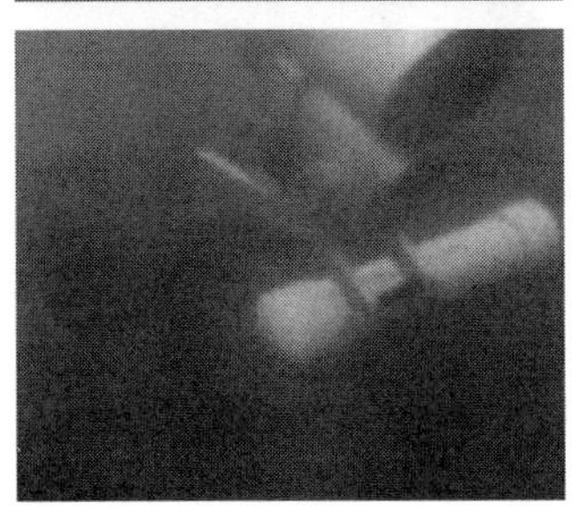

FIGURE 9.1: This example shows a typical error that occurred while the operator was attempting to control the remote manipulator to retrieve a capsule from the sea floor. The depicted image fragments were automatically selected by the system (see Section 8.4), compressed, and transmitted to the operator station during task performance. In this case, the gripper failed to fully close on the object. When the slave detected this departure from expected motion, it signaled the master station. The operator could then examine the received imagery to diagnose the problem.

with the task. Since the system is now aware that the hole is not empty, it can again aid the operator in attempting to remove that bolt.

The operator also has the ability to change the "conservativeness" of the generated commands. For example, when unscrewing a bolt it is possible to double-check that the bolt was correctly extracted by tapping it on a nearby surface. When a high level of conservativeness is chosen, the system will automatically insert commands for "tapping" whenever an unscrew operation is performed. This costs bandwidth when the commands are transmitted and time when the commands are executed, but it saves time in those cases where unbolting is unsuccessful by detecting them as soon as they occur. By allowing the operator control over this aspect of command generation, we allow the intelligence and experience of the operator some influence on the form of the generated commands.

There is a tradeoff in deciding how much, and in what form, information should be provided to the operator. If too little information is provided, then the operator may incorrectly diagnose the exceptional condition. This wastes time since failing to correctly rectify one problem invariably leads to others occurring later. Alternatively, if the operator is provided with too much information, then bandwidth and time is wasted moving that data from the slave to the master station. It is expected that the optimal information display will vary depending on both the particular task being performed and the particular human operator performing that task.

9.7 Summary

Practical implementations must include some consideration for exceptional events. The synthetic fixtures introduced earlier in the book serve to help guide the operator's actions. This allows the system to both guide the operator around uncertainty and to unambiguously interpret his or her actions.

It is unrealistic to expect that all errors may be avoided. In the teleprogramming system, errors are handled by using a three stage process: autonomous detection, shared diagnosis, and manual recovery. We believe this is an appropriate use of technology. We rely on the operator handling the complex reasoning required to recover from errors while automating the tedious and time-consuming task of observing the slave's actions looking for errors.

10

Command Generation and Interpretation

"Computers are good at following instructions but not at reading your mind."

—*The TEXbook, Donald E. Knuth, Addison-Wesley, 1993.*

This chapter describes the implementation of the teleprogramming language used to carry symbolic information between the master and slave sites. In common with many existing robot programming languages (see for example [59, 48, 18]), we had the usual requirements of performing relative or absolute joint or Cartesian-space motions while allowing detected sensory signals to change the flow of program execution. In addition, the limited communications channel and uncertainty in the world model acted to constrain the symbolic telemetry.

In this, the second implementation of a teleprogramming language, we move away from hybrid control, implement conditional expressions, allow motions relative to symbolically defined frames, and define a language for replies from slave to master. The language includes support for visual imagery (see Chapter 8) and error recovery (see Chapter 9). A full definition of the implementation is given in Appendix A.

10.1 Master-to-slave teleprogramming language

Each generated command is composed of five sections:

1. a beginning
2. pre-motion instructions
3. motion instructions
4. post-motion instructions
5. an end

For example, here is a command to move the manipulator to the zero position and then assign that end-effector position a symbolic frame f1.

```
B 5
C c0 220 240 1 0.2
C c1 310 430 2 0.5
M j0=0
M j1=0
M j2=0
M j3=0
M j4=0
M j5=0
D f1=ee
E
```

The beginning and ending statements are important. The beginning marker encodes a unique identifier for the enclosed command. These allow the master and slave stations to refer unambiguously to particular commands. The end statement advises the slave that the whole command has been read, thus enabling it to begin execution.

In this example, the pre-motion commands encode specifications for the visual imagery system. For each of the two cameras (c0 and c1), they describe the preferred region of the image by specifying the center location (in pixels) and a subsampling ratio. They also assign a score to each camera based on the perceived usefulness of its view. Intended to be generated and interpreted by computer, these instructions are not particularly human-friendly.

The motion commands describe the action to be performed by the slave robot, in this case, specifying a joint space motion to the zero position.

The post-motion commands encode instructions to be performed after the commanded motion has terminated. In the above example, the post-motion instruction assigns a symbolic label (f1) to the position of the end-effector (ee) at the end of the motion. Subsequent Cartesian-space motions may be specified relative to that symbolic location.

10.1.1 The need for motion relative to absolute frames

One approach to the generation of Cartesian commands would be to describe each motion based on the absolute position of the slave end-effector. This is very simple to generate and avoids any problems with drift during task performance. The disadvantage, however, is that any discrepancy between the real and modeled world will result in positional errors—it is not possible for the slave to adapt to its environment without changing the master station model. While this would certainly be possible, it comes at a cost since information about the slave site takes at least one communications delay to reach the master station, and it will be at least another communications delay before any new commands incorporating that information can reach the slave site. Thus, any adaptation to the slave site will lag at least one whole round-trip delay time behind the slave's motion.

An alternative approach is to specify all motion as relative to the current end-effector position at the slave. This allows the slave to adapt to the environment. For example, if the slave is commanded to come into contact with a surface and then move 2 cm away from it, then it will end up 2 cm away from the real surface at the slave site rather than wherever the master station model predicted that position to be. These simple relative motions allow the slave to accommodate for modeling errors without the need for any model updating. The difficulty with this approach is that any positioning errors become compounded. For example, if each motion is 1 mm too long and the slave makes five motions in the same direction, then it will be 5 mm off and, more importantly, it will not "know" that its position is in error.

A third approach, and the one currently employed, is to assign symbolic labels to features in the world and then specify motions relative to those features. This allows the slave to adapt to the real world while avoiding the problem of positional drift. It also gives the slave some memory of the remote site, allowing it to make use of the knowledge gained from previous contact with the world. This approach only requires that the world model be locally correct.

10.1.2 Defining frames

The system assigns symbolic labels to features in the world by defining frames either at, or relative to, the current end-effector position. Each new frame requires memory to store it at both the master and slave stations. Also, as the number of frames increases, the time taken to locate any particular frame also increases. As a result, it is undesirable to define a new frame at the end of every motion—we need a more intelligent method.

If we assume the world never changes, then there is no need to define a frame more than once in the same location. Nor is there any need to define a frame unless it is significantly different to existing frames. In this context, we define "significantly different" as being either spatially different (in the current implementation at least 100 mm away from a previous frame), or topologically different (the end-effector has different surfaces in contact with surfaces in the world). In cases where we allow for variation in the model over time, then two frames are also different if they have sufficient temporal separation.

The system defines frames relative to other frames, building up a tree of symbolic labels with respect to which it can define subsequent motions in the world (see Figure 10.1). The use of a tree structure is intended to allow for future consideration of uncertainty. The uncertainty associated with each frame depends on the uncertainty of its "parent" frame. If one frame is found to be very inaccurate, then all of its "children" are also likely to be inaccurate.

When an error occurs, any frame(s) defined during or after the command in which the error occurred are deleted (the commands in which they were defined were never successfully executed by the slave).

10.1.3 Selecting frames

At present, the selection of an appropriate reference frame is made by giving each frame a score and then choosing the frame with the lowest score. This score is evaluated by considering the distance of the end-effector from the

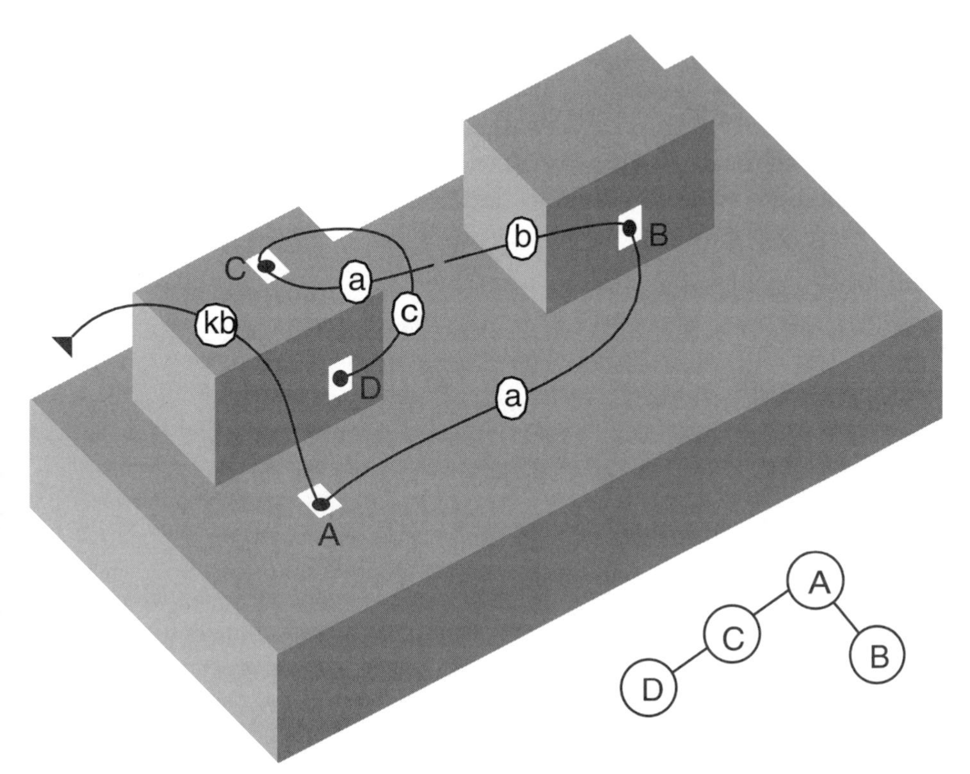

FIGURE 10.1: Simple example motion sequence showing locations where frames are defined. The labels on each path segment indicate the frame with respect to which that motion was specified (since there are initially no frames defined, the first motion into contact is specified relative to the kinematic base frame, KB, for the robot). Note the motion between points B and C begins by using frame B as reference and then switches to frame A. Also shown is the tree of frames resulting from this motion.

frame, the number of contact points with the environment when that frame was defined, and the time since that frame was defined. This score is an *ad hoc* measure of uncertainty.

It is interesting that the system may command the slave effectively, despite the fact that there is almost a complete separation of knowledge between the two stations. At the time commands are generated, the master station does not know the exact location in the world of the frame with respect to which it is generating commands. While the slave site does not know what feature each frame refers to—it just knows that they are places it has been to before.

10.1.4 The frequency of command transmission

In generating commands, one option would be to send a large number of very small commands. This approaches conventional teleoperation where joint information may be transmitted at rates between 100 and 1000 messages per second. An alternative would be to send messages very infrequently, perhaps as few as one per hour. This approaches fully autonomous operation where there is theoretically no required communication during task performance. In practice, neither of these extremes is desirable or possible.

Sending many small commands is inefficient for many communications links, since there is a fixed overhead associated with each transmitted packet. Sending few large commands is also undesirable. As command frequency is reduced, the duration of execution described by each command must increase. This hampers command generation by requiring increasingly higher levels of representation and complicates command execution by requiring the slave to operate autonomously for increasingly long periods of time.

We compromise by sending commands at variable rates, on average one or two commands each second. This allows each command to be small enough that it is simple to generate and execute while still maintaining a relatively low command transmission rate.

10.1.5 *Handling the communication time delay*

The communications time delay between sites makes it undesirable to have the generation of the current command dependent on knowing the result of execution of any previous command. This poses a problem since the operator station must generate commands without knowing what may occur during the delay between command generation at the operator station and execution at the slave. This is handled in three ways.

The first method is to allow actions to be specified that are functions of previously measured sensory data. This is achieved by specifying motions relative to symbolic frames (as described above). At the operator station, the modeled slave manipulator is moved relative to the expected location of those features, while at the slave site the real robot executes motions relative to the actual measured feature locations.

The second method is to recognize that there will be cases where there are a small number of expected slave states and the desired final result of execution in each state is the same. In this situation, the operator station may send several parallel execution streams (one for each possible state) to the slave site and allow the system to switch between those streams during execution depending on its sensed state. Since all alternatives converge to the same desired state, the master station may continue to generate commands for subsequent actions without needing to "know" the particular path by which the slave manipulator will reach that state.

The final method is to detect cases where execution does not match that anticipated by the operator station. Expected sensory readings are encoded within the transmitted commands and the slave may compare those expectations against real measured data to detect cases where reality departs significantly from the operator station model.

These latter two methods require instructions to allow conditional expressions and branching based on measured sensory data.

10.1.6 *Sensors and conditional expressions*

The current implementation has sensors for joint limits, joint torque limits and gripper contact. While these could refer to real physical sensors, in most cases they are constructed entirely in software. For example, the torque limit switches are activated whenever the computed torque required from any motor exceeds the software-encoded limit for that motor. In addition there is a virtual sensor (called s0) that is activated if, and only if, the most recently commanded motion ran to completion.

The conditional expressions compare a sensor value with an expected value and, if the condition is true, may either generate an error or step forward any number of commands.

For example, this code fragment closes the gripper (which happens to be joint 5 on this manipulator) and tests the gripper contact sensor (which happens to be sensor 1000).

```
B 6
M j5=0.3
I s1000=1 1
I s1000=0 0
E
```

Upon receiving this command, the slave robot will attempt to move joint 5 to a position of 0.3 radians. It will perform that motion until any observed sensor changes state. In this case, since sensor 1000 is explicitly mentioned in the post-motion commands. The slave will observe that sensor while closing. Thus, when the motion terminates, there are two obvious possibilities: either the motion terminated because it reached its natural conclusion (the gripper closed all the way to the 0.3 radians point and the virtual sensor s0 activated), or alternatively the motion was stopped prematurely because the gripper contact sensor changed state.

There are two instructions to test for these possibilities. If the contact sensor is true, then the interpreter is instructed to skip forward 1 command; continuing with the very next command in the instruction stream. Alternatively, if that sensor is false, then the slave should treat that as an error and remain at this point (effectively skipping 0 commands) until new post-error commands arrive from the operator station.

Now, while those are the two obvious possibilities, there are in fact many other possible errors. For example, the act of closing the gripper could cause some other joint to exceed a torque limit. Rather than explicitly coding for every possible eventuality, the interpreter assumes a number of implicit conditional statements. Specifically, it implements a policy of self-preservation, automatically predicating each motion on any sensor that warns of possible danger to the manipulator.

10.1.7 *Moving within the command stream*

In developing the system, a conscious decision was made to never allow the slave to jump backwards in the command stream. Thus, each command sent to the remote site is executed at most one time. This has the following advantages:

- It keeps the slave site simple by avoiding any need for the slave to store previously executed commands.
- Since all commanded motions have finite duration, and since each command can be executed at most once, there is no possibility of the slave entering an "infinite loop." To check for possible software coding errors, it would be practical to implement a watchdog timer to check that (so long as commands continued to arrive from the operator station) a new command was being executed at least once every t seconds. Where t is the maximum time for any programmed motion.
- It simplifies error recovery since the unique identifiers for each generated command also provide unique identifiers for each executed command.

The cost of not allowing negative branches is that, if actions are to be repeated, the operator station must encode and send those actions multiple times.

10.1.8 *Command conservativeness*

On the one hand, we could send nothing but very low-level commands to the slave (this would approach conventional teleoperation); on the other hand, we could send very high-level commands to the slave (this would approach automation). In reality, we want a system somewhere between these two extremes. Some autonomy is necessary and desirable to cope with cases where local high-bandwidth feedback is required. As more autonomy is added to the slave site, it becomes easier for the operator to command the slave but it also becomes much harder for the operator to figure out what has gone wrong when an error occurs. If the transmitted commands are relatively simple, then the operator gains a lot of information from merely knowing in which command the error was detected. On the other hand, as transmitted commands become more complicated, the number of different things that can go wrong increases greatly and error diagnosis is problematic. So, at least in terms of error recovery, we would like to be transmitting and executing relatively simple commands.

The difficulty with this approach is that it requires a larger bandwidth and it requires the operator to repeat very similar motion sequences again and again. For example if, each time a bolt is unbolted, the operator would like to tap it on the surface (to ensure that it has been successfully unbolted) then he or she must manually perform the tap after each unbolting. This becomes tedious, but it forces the operator to be aware of exactly what the slave will try to do and hence makes it easier for the operator to recover from exceptional events.

One way to work around this is to hard-code the required sequence of commands; for example, whenever the operator performs an unbolt, the system could send commands to unbolt and then tap the surface with the bolt. The system currently implements this policy. In addition, it animates the generated commands for the operator (so he or she is aware of what commands have been generated) and gives the operator some control over the generated command sequence. The operator can select between different command generation modes ranging from very conservative to very fast—conservative modes generate commands that double-check operations by, for example, tapping a bolt to make sure it has actually been extracted. Fast modes generate commands to do the minimum possible amount of work with no additional testing. The operator has the option of changing modes at any time and will generally do so in response to the success or failure of commands at the remote site. If the world model matches well with the real world and few errors are being generated, then a faster mode is better. Alternatively, if errors are occurring frequently, then a conservative mode is preferable—more commands will be generated and execution will be slower, but it will be much easier to detect/diagnose exceptional events.

10.2 Slave-to-master teleprogramming language

The slave-to-master portion of the language is primarily dedicated to initialization and error detection and recovery.

10.2.1 Initialization

The teleprogramming language must provide a mechanism for restarting the system from an unknown state. This is particularly important and often ignored. In the laboratory, it is simple to move the manipulator to a known position and configuration prior to performing an experiment. However, with the teleprogramming system the remote site will often be inaccessible to the operator and environmental conditions could make conventional visual feedback unavailable. Thus, there is a need for slave-to-master communication at the beginning of any new task session to transfer a symbolic description of current slave state to the operator station.

This is achieved by having the slave send a copy of all available joint positions and other sensory information within the first message transmitted to the master station.

10.2.2 Command replies

As the slave performs each commanded action, it records joint and sensor information at relatively high rates (it is conceivable that rates in excess of 1000 hertz could be achieved). Sending all this recorded data back to the operator station is impractical due to bandwidth restrictions, and unwise since much of it will never be used by the operator.

Thus, the current implementation adopts a multiphase process. It compromises between the cost of sending information to the master station and the indirect cost of attempting to diagnose errors with insufficient information.

As the slave completes each commanded action, it sends a single, relatively simple, reply to the operator station. This message encodes an identifier for the command just completed, along with a timestamp, and the current slave state. For example, the following reply was sent at time 32.5 after completion of the example move-to-zero-position command given above.

```
E 5 32.5 0 0 0 0 0 0
```

These messages are used to create informational displays for the operator. (For example, the master station displays an estimate of the delay throughout the system derived from the difference between the time each command was generated and the time a reply was received.) They are also cached at the operator station. This cache contains a coarse record of past slave states that may be used to create approximate simulations of prior slave actions.

When the slave detects an error, it sends a series of messages to the operator station. These begin with a description of the error and the current slave state. This is followed by a historical record of recorded slave motion for several seconds prior to the error. This additional data fills in the gaps between the relatively coarse state information previously transmitted in the command replies. Since the amount of data may be large, it is sent in reverse time order. Thus, the most recent (and presumably most useful) data is sent first. The operator may view simulations of prior slave actions as soon as the error message begins to arrive, and those simulations become more detailed and cover a longer time history as the following messages are received.

10.3 Delaying command execution

In the current system, the slave executes each received command as soon as it is able. An interesting possibility for future implementations is to deliberately delay commands at the slave site before execution. The advantage is that, by knowing which commands were required next, the slave may be able to make better use of available resources. For example, it could schedule when images should be taken by considering the requirements for successive commands. In situations where both a remote vehicle and manipulator were being controlled, it could reposition the vehicle to optimize its position in preparation for subsequent motion commands.

10.4 Adding additional sensory feedback

In the present implementation, symbolic frames are defined at the slave site using only information from joint sensors. There is no reason why other sensors could not also be employed. The use of visual imagery to track features in the remote environment should be quite feasible (in fact, it has already been used for visual station-keeping underwater [50] and has been suggested for use in space telerobotics [32]).

For tasks such as object retrieval, we envisage a system that works in the following way:

1. The slave system looks for features in the images, assigns symbolic labels to any that are found, transmits that information to the operator station, and begins to track the features in the image plane.

2. The operator indicates corresponding features among the set found by the slave. This allows the 2-D image-plane positions to be converted into 3-D locations. The labels of this correct set are transmitted to the slave.

3. Upon receiving the master station reply, the slave continues to track only those that were confirmed by the operator, while generating symbolic positions from those features for use in executing subsequent commands.

4. When generating commands for subsequent motions the operator station may make use of its model to predict the expected location of the features and then specify motions relative to the corresponding symbolic labels. This is analogous to the current implementation that allows motions to be specified relative to locations the slave has previously contacted. Since the slave may not be able to successfully track all labeled features, there may be an advantage in having the master station generate several alternative motion commands and allow the slave to choose to execute the one for which the tracked features have the least uncertainty.

10.5 Summary

The teleprogramming language is used to convey information between sites. For commands sent from master to slave, it allows motions relative to symbolic frames as well as conditional expressions and an ability to branch forward in the command stream. In addition, the system supports the notion of command conservativeness, allowing operators to trade off between fast and risky or slow and safe command sequences.

The replies sent from slave to master facilitate initialization and error diagnosis, providing the operator station with enough information to allow it to simulate prior slave actions.

When transmitting information between sites, there is a continual need to trade off between the small certain cost of sending additional information, and the large possible cost of having insufficient information to avert, or quickly recover from, an unexpected event.

11

Results and Observations

This chapter presents results and observations obtained using a number of different test systems. The initial experiments, conducted under contrived conditions both in air and in a test tank, were intended to validate the software and hardware employed. The idea was to push the boundaries of the operational envelope, finding and fixing systematic design weaknesses. The later experiments, intended to test the feasibility of the approach, were performed in the sea under real conditions. For the October experiments, the operator was located in Woods Hole, Massachusetts, around 100 meters from the submerged slave device. While in the final, November experiments, the operator was over 500 kilometers away in Philadelphia, Pennsylvania.

11.1 Laboratory trials

The initial implementation of the teleprogramming system was in air (see Figures 11.1 and 11.2) and did not provide any visual feedback from the real slave site. The operator and slave device were separated by approximately 10 meters and a communications delay of 10 seconds was implemented in software.

The master station provided the operator with both visual and kinesthetic feedback based on a model of the remote environment. Processing was split

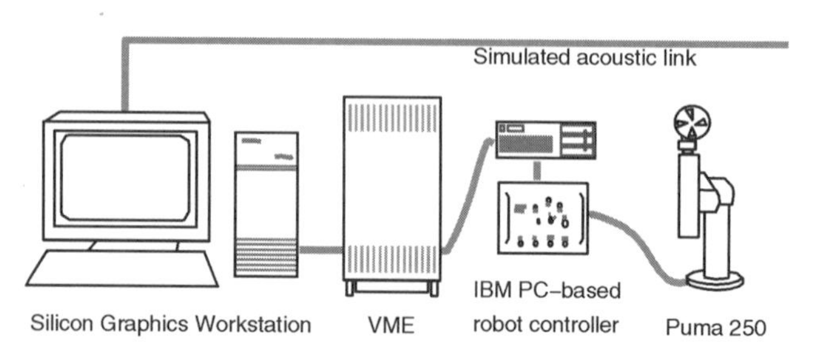

FIGURE 11.1: In this experimental implementation, a Silicon Graphics workstation was used to maintain the world model and display the virtual world to the operator at the master station. The operator commanded the system by moving the end of a small Puma 260 robot.

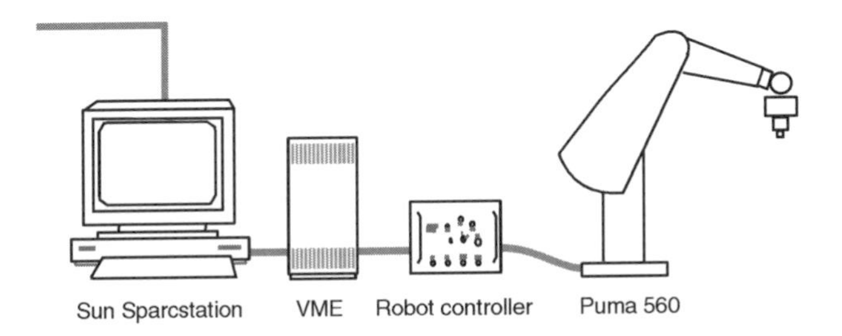

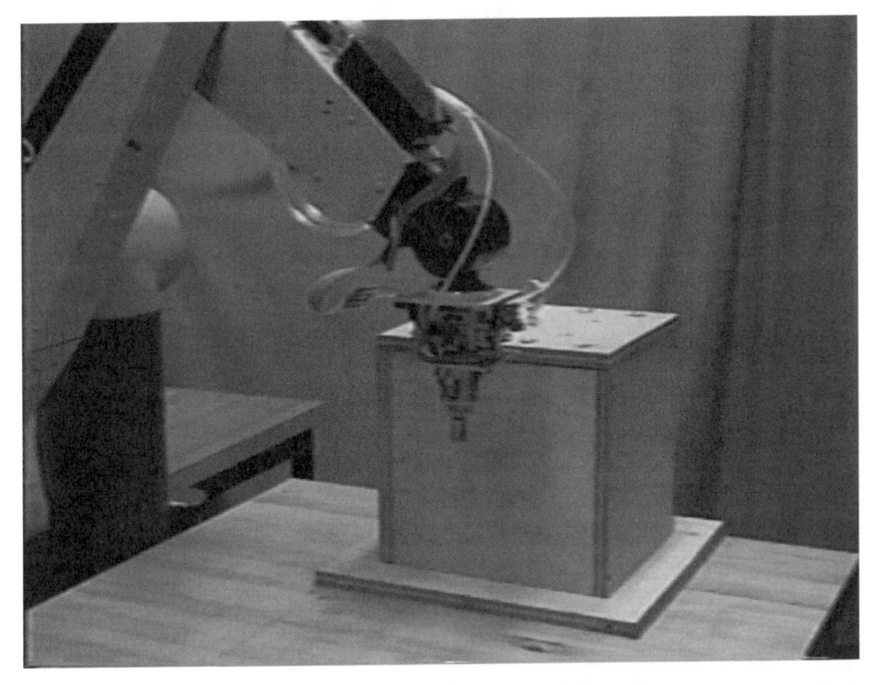

FIGURE 11.2: The slave station utilized an industrial Puma 560 robot equipped with an instrumented compliant wrist. Communication between sites was via a UNIX socket connection and was artificially delayed to simulate a distant remote site.

between the graphics workstation and a high-speed coprocessor [4]. The graphics machine determined which fixtures should be active and displayed the virtual world, along with appropriate visual clues, to the operator with an update rate of approximately 12 Hz. Data for active fixtures was passed to the coprocessor which combined this with measured operator-applied forces to generate appropriate motions for the master arm.

The slave station for these in-air experiments was developed by Lindsay [47] and Stein [79]. It interpreted the teleprogramming instructions, performing commanded actions using a Puma 560 industrial robot equipped with an instrumented compliant wrist. A behavior-based controller provided a low level of remote site autonomy, allowing the slave to react immediately to the sensed deflection of the compliant wrist.

Using this system, it was possible to perform tasks ranging from simple contact operations [30] to more complex operations, such as bolting and unbolting a hatch cover [72].

11.1.1 Example task

The example task is to remove and replace the bolts on a hatch cover. In this first implementation, there are no sensors at the remote site other than the robot joint positions and the instrumented compliance in the wrist. Lacking sufficient sensors to build a world model, it is assumed that the operator station has *a priori* an approximate model of the remote site, but that model is not assumed to be perfect.

The master station provided the operator with a virtual reality representation of the remote environment, which included both visual and kinesthetic feedback. As the operator worked, the system continuously observed his or her actions and translated them into a sequence of symbolic commands for transmission to the remote site.

He or she performed the task entirely within the simulated world. The operator's interaction with the system is through a natural teleoperation-style interface that supports reindexing and provides scaling and kinesthetic correspondence (see Chapter 6).

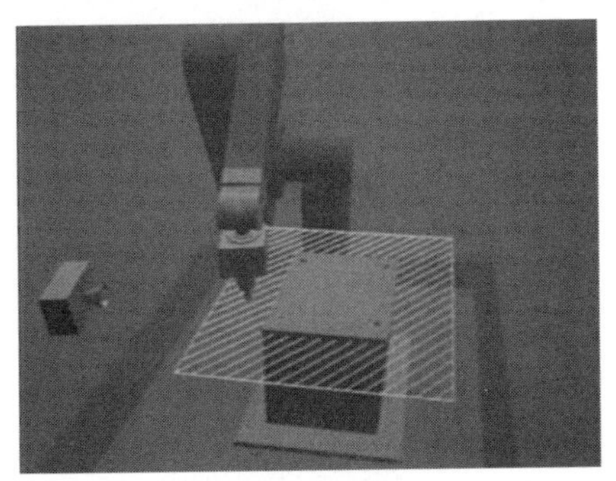

The operator begins the task by moving to contact the box. The system will assign symbolic labels to those contact locations and describe subsequent motions relative to them (see Chapter 10). This allows the operator station to generate commands without needing to wait for the first replies from the remote site.

As the operator works, the system attempts to predict his or her next action and activates appropriate fixtures to assist in task performance. Here, there are four active fixtures. The two planar fixtures assist in contacting the box, while linear and command fixtures aid in bolt removal (see Chapter 7).

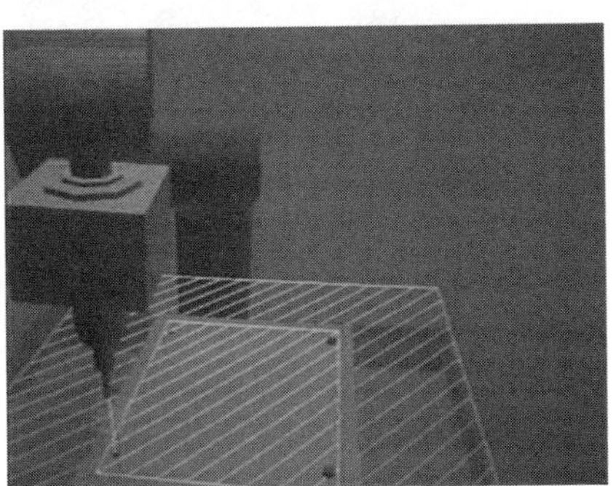

By pressing though the command fixture, the operator unambiguously commands the system to remove the bolt. Then, as he or she moves the end-effector forward, the end-effector nears another bolt hole and the master station activates a different linear fixture to guide the impact wrench over that location.

Meanwhile, at the remote site the slave robot has already begun to receive and execute the first commands (see Chapter 5). Each performs a single action and incorporates some knowledge of the expected result. For example, here the slave performs a motion and expects to contact the box.

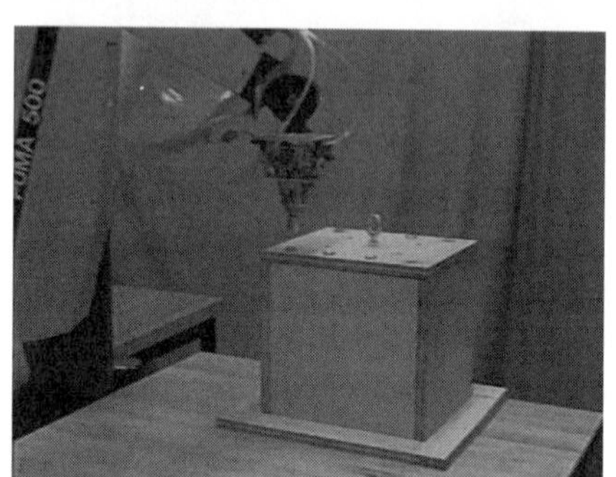

Having contacted the box, the slave is able to replace the symbolic positions in the received teleprogramming commands with actual measured locations. Thus, it is able to repeat the operator's motion, moving over the bolt, even though the box location may differ from that in the master station's world model.

Having successfully extracted the bolt, the system executes subsequent commands, which instruct it to tap the bolt on the box (to test for its presence in the impact wrench). Provided the test succeeds, it is able to continue with the task.

The bolt insertion operation was encoded and transmitted to the slave as a sequence of around twenty simple commands. These instruct the slave to repeatedly move down into contact with the surface around the expected hole position and then, when the hole is located, activate the wrench to drive the bolt in.

11.1.2 Observations

While tasks could be successfully completed, it was very difficult for operators to recover when things did not work as planned. What made the problem hard was that, particularly in the bolting/unbolting demonstration, the slave site had state. Not only did the operator have to determine what had gone wrong, but he or she also had to determine the current location of each object at the slave site. Without some form of panoramic sensor (such as vision), the operator was left with no choice but to make a series of exploratory motions, testing and eliminating different alternatives. As a result of experiences with this system, it was considered essential that a panoramic sensor be employed at the slave site and that data from that sensor be made available to the operator.

11.2 Test-tank trials

For the test tank trials, a new slave system was developed using the manipulator from the JASON subsea vehicle (see Figure 11.3). Migrating to this system involved much more than just a change in the appearance of the slave manipulator. A new teleprogramming command language was developed. The slave system was now position-controlled. A slave command interpreter was developed and real remote site cameras were integrated into the system. Details of the implementation are given in Appendix A.

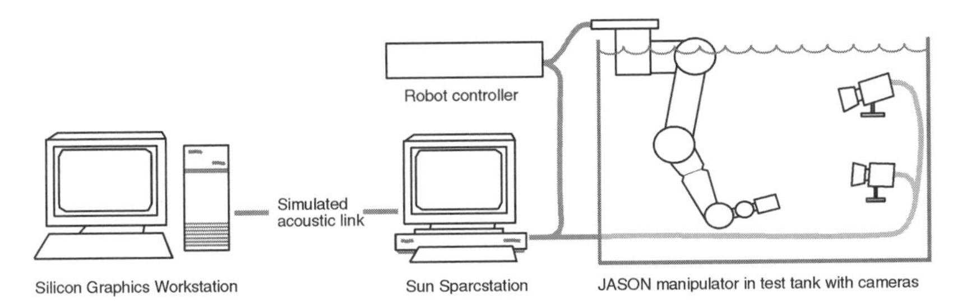

FIGURE 11.3. The test-tank implementation.

The command interpreter acted as an intermediary between the symbolic commands transferred between master and slave sites and the low-level joint and sensor information available from the JASON arm controller. This division between high- and low-level functionality was physical, as well as logical. The higher-level functions were implemented on a Sun workstation with a serial link to the arm controller in its titanium pressure housing.

Images from the remote cameras were digitized in the Sun workstation under control of the command interpreter. As described in Chapter 8, these images were used to calibrate the cameras, maintain the world model and aid in situational awareness. There was now no *a priori* assumption about the position of objects at the remote site, and panoramic feedback was available to assist in error diagnosis and recovery.

At the master station, provision was made for controlling the simulated slave robot using a mouse. Cartesian control was implemented by allowing the operator to simply click and drag the displayed end-effector. Joint space motions were made possible using a set of on-screen sliders.

For these trials, the operator was located in a separate room from the test tank (far enough away so he or she could not hear what was happening), and communication between master and slave sites was delayed in software to simulate a 10 second round-trip delay time. A flask was dropped into the tank and the example task was to locate and retrieve it (see Figures 11.4 and 11.5).

11.2.1 Example task

The example task is to retrieve a flask. The master station has a graphical model of the arm and the flask. It does not know the position/orientation of either the flask or the remote site cameras.

Camera calibration

The operator's first task is to calibrate the cameras. This is achieved by using the remote manipulator as a known reference object. The manipulator is, under teleprogramming control, moved to a number of positions while images are periodically recorded. The operator then manually indicates correspondences between features in those images and the same features on the robot. This information is sufficient to enable determination of camera parameters. Position and orientation are determined with respect to the kinematic base frame of the robot.

World model updating

The operator then proceeds to update the graphical model of the remote environment. The real images of the remote site are first overlayed on top

FIGURE 11.4. A typical operator display during test tank-trials.

FIGURE 11.5: Image from a camera submerged in the tank. The shown flask was successfully retrieved from both vertical and horizontal positions.

of the simulation. The operator may then manually correct any discrepancy by dragging and stretching the modeled objects. This interface is fast and natural. In this case, where the object is of known size, it only takes a few seconds for the operator to drag the modeled object to the correct location.

Commanding the remote robot

The remote manipulator may then be moved into position to grasp the test object. The operator does this by moving the simulated robot (either by dragging the end of the robot with a mouse or by using the master manipulator). Just as in the in-air tests, the system observes the operator's interaction and continuously translates those motions into relatively high-level teleprogramming commands for transmission to the remote site. The difference is that now those commands also include the necessary information for automatic control of remote site cameras. They support intelligent fragmentation, as well as control over both the frame rate and rate of task performance (see Chapter 8).

Once again, synthetic fixtures aid the operator's interaction with the system. In this example, they aid the operator in aligning the gripper with the object to be grasped. As the operator approaches the flask, a line fixture is activated. This pulls the operator toward a perfect approach trajectory. As the gripper encompasses the object, a command fixture activates. This is, in effect, a virtual surface between the object and the gripper. By pressing through this surface the operator unambiguously indicates his or her intention to pick up the object and the system automatically generates appropriate gripper closure commands.

11.2.2 Command execution

The slave robot executes each command as it is received and compares measured performance with that expected by the operator station. For example, the command for closing the gripper includes the expectation that the gripper encounter an object before reaching a motion limit. If this expectation is met, then the slave may simply continue to execute subsequent commands. If an unexpected sensor reading is detected, then the slave pauses execution, sends a signal to the operator, and awaits the arrival of new post-error commands.

11.2.3 Handling execution errors

When the operator station receives an error message, it "winds back time" to the point where the error occurred. The slave system's preliminary error diagnosis is then displayed for the operator. In this example task a typical message is "gripper contact not detected." The operator then has the option of stepping either forward or back through the historical record of previously executed commands. For each command, displays are available of expected and recorded slave motion as well as of images taken while those motions were being performed.

Using this information, the operator must diagnose and recover from the error. This may involve editing the world model to bring it into closer alignment with the real world. It may also involve correcting any invalid assumptions the system may have made concerning the success or failure of past commands. The operator may then continue with the task, generating new commands for the remote robot.

11.2.4 Observations

During initial operations, the fourth joint of the arm (controlling wrist rotation) failed and only a single camera was operational. However, even with these hardware failures, task performance was still possible. The presence of only one camera was mitigated by the presence of a flat surface within the test tank as well as by the *a priori* knowledge of the size of the object. The operator was able to manually compensate for the joint failure by avoiding Cartesian motions that would have required large motions of the failed joint.

The operator utilized high resolution images from the remote site to locate the test object and then, with those images overlayed on top of the simulation, the simulated test object could be dragged across the screen until its position matched that of the real object. Within the simulation, the slave robot was then moved into position so that its gripper enclosed the object. Activation of the gripper closure action was then initiated with aid from a command fixture.

The generated program commands for this maneuver incorporated the expectation that the gripper encounter an object while closing. When the slave executed those commands, it monitored the commanded torque for the gripper motor. If the expected sensor reading occurred, then the slave continued executing subsequent commands. If it was not detected, then the slave paused execution and sent an error message to the operator station.

It was found during testing that the gripper contact sensor was too sensitive and the slave would often "think" that it had grasped the object before it was securely held. This meant that the flask sometimes slipped from the gripper. The slave had no way to sense this and so continued executing subsequent commands. It was, however, very obvious to the operator when viewing the delayed visual imagery. Thus, while it wasted time (the operator had to repeat part of the task), it did not prevent the task from being performed. This is an example of a situation where the slave's autonomous sensing ability was inadequate, yet the presence of the operator meant that task performance was still possible—albeit at a reduced rate.

The JASON manipulator has only five degrees of freedom (it has no equivalent of the joint 6 rotation on the Puma robots). This, combined with the failure of the wrist-rotate joint, meant that the arm had insufficient dexterity to reach every possible object position. In particular, it was not possible to directly retrieve the flask when it was lying on its side with its axis orthogonal to the direction in which the arm was reaching. In a real subsea situation, the solution to such a problem would be to simply move the vehicle, reorienting the arm with respect to the flask. Clearly this was not practical in the test tank. However, the operator was still able to perform the task. The solution was to use the arm to push the flask, rotating it by approximately 90 degrees and, in effect, reducing the retrieval operation to a previously solved problem.

It was during these trials that the operator was, for the first time, provided with real images from the remote site. While this was certainly required for task performance (it was the only way to determine flask position), the presence of real visual imagery had a more subtle psychological influence on the operator. When working with the in-air system, the lack of any visual feedback left operators with a continual nagging doubt as to what was happening at the slave site. The slave might not be reporting any errors, but its ability to detect errors was imperfect—there was never any guarantee that everything was working perfectly. The presence of even occasional images from the remote site alleviated much of this doubt, leaving the operator feeling much more confident that he or she was in control. It was also much more satisfying. In cases where task performance became impossible (such as the flask rolling out of the robot's workspace), this was apparent to the operator who could then just stop. There was never any sense of wasted effort.

During these trials, there were three significant failures (one camera, one joint, and one sensor). Yet, despite these, the task was still performed successfully. This bodes well for the robustness of the approach.

11.3 Migrating to a subsea system

For the move to a subsea system, the Sun workstation was mounted in the JASON control van (see Figure 11.6), and the manipulator was reinstalled on the JASON vehicle [94] (see Figure 11.7). The serial link between the Sun and arm control computers (now returned to their pressure housing inside JASON) would now be via the optical fibre cable inside JASON's tether. Video signals from the cameras mounted on JASON were similarly transmitted to the surface. The JASON arm could now be controlled either manually (from the button-box and joystick located in the JASON control van) or remotely (via the teleprogramming link). An overview of the implementation is shown in Figure 11.8.

During the teleprogramming experiments, weight was added to JASON to make it negatively buoyant (it is normally approximately neutral), and it was piloted to the seabed. Vertical thrusters were then continuously employed to keep the vehicle settled on the bottom while the arm was deployed, calibrated, and the teleprogramming experiments begun.

The chosen task was the retrieval of a flask dropped on the sea floor. Prior to each test trial, the manipulator (under manual control) was used to remove a flask from a basket on the vehicle and drop it in front of the vehicle. The arm was then moved away from the flask before control was turned over to the teleprogramming operator. His/her task was then to locate and retrieve the flask. Once the trial was complete, control was returned to the manual operator and the drop-retrieve cycle was repeated.

The arm that was used to set up the experiment was the same arm that was used in the experiment. This introduces some bias since it makes it more likely that the object will be favorably oriented. However, this bias is consistent with that which would have been present if the pilot had had the opportunity to reposition the vehicle prior to each retrieval trial.

FIGURE 11.6: The JASON control van. The manipulator may be controlled from the engineer's station at left, while the vehicle is controlled from the pilot's station at the right. During operations at sea, a third station to the right of the pilot is used to support navigation and liaison with the ship's bridge.

FIGURE 11.7: The JASON vehicle as configured for the teleprogramming experiments. Note the presence of the manipulator arm (in stowed position). The dark cylinders pointing toward the arm are pressure housings for the subsea cameras.

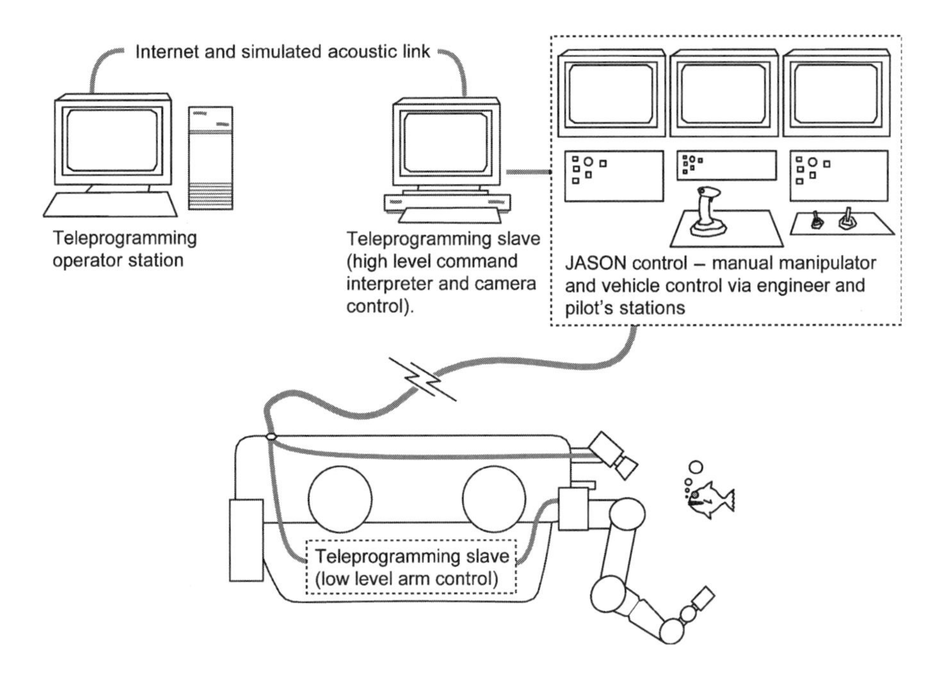

FIGURE 11.8. The subsea test system.

11.4 The October experiments

For the October experiments, the teleprogramming operator station was located in the JASON control van around 100 meters from the submerged vehicle, and a 10 second transmission time delay was again implemented in software.

The system worked. The chosen task of retrieving a flask was successfully performed over a dozen times. Task completion times varied over a wide range—from less than three minutes to nearly an hour. The differences between execution times becomes clearer if we divide the task into stages:

1. Model update: During this stage, the operator overlays views of both cameras on the graphical model and then proceeds to update the model by dragging and rotating the graphical image of the test object until it matches the imaged position of the real object.

2. Move to grasp: In this phase, the operator uses a mouse to command the simulated robot to grasp the flask.

3. Achieve successful grasp: This stage covers all actions from the end of stage 2 to the completion of the task. Completion is defined as that

time when the operator is convinced (from real visual images of the remote site) that a successful retrieval has been achieved.

Updating the model consistently took less than 1 minute. The fastest time was 25 seconds, the slowest 55 seconds. This level of performance makes it difficult to justify moving to a more automated technique (particularly given the poor quality of the images and variability of the environment and lighting conditions).

For moving to grasp, the fastest time was 1:12 while the slowest was 2:09. The variation in times was largely due to operator inexperience. Limitations on joint ranges and available degrees of freedom meant that a single arm position was inappropriate for all attempted grasps, and the operator would often experiment with several different arm configurations before finding a good position. The typical strategy was to perform several joint space motions to move near the object while adopting the desired configuration and then move to Cartesian-space motions to complete the grasp. It was often necessary to rotate the world view in order to position the gripper along all desired axes.

Achieving a successful grasp showed by far the most variability, with completion times varying from under 20 seconds to over 20 minutes. The variability here depended largely on the success or failure of the physical grasp. If the initial grasp was successful, then a completion time of less than 20 seconds was achievable. If the grasp failed, then the operator had to diagnose the problem, reposition the gripper, and retry the grasp.

Some situations were particularly difficult to diagnose. In one case the uneven terrain meant that the far side of the gripper (which was not visible in either camera view) contacted the sediment layer before touching the test object. Coincidentally, a small marine plant was caught between the front of the gripper and the object. Thus, when the grasp failed, the operator incorrectly diagnosed the problem as being caused by the plant and wasted much time needlessly trying to avoid it before finally obtaining a successful grasp. In this case, better visual imagery of the grasping action would have alleviated the problem.

It was found that grasping was very sensitive to the relative positions of gripper and object. A particular difficulty was that any contact with any finger was sufficient to trigger the "grasp-completed" sensor. Thus, successful grasps were signaled even though it was often the case that none had been achieved. The operator quickly came to realize that the local grasp sensor was unreliable and moved to the more conservative strategy of double-checking grasps by attempting to further close the gripper and confirming closure using visual feedback. Many of the grasping problems encountered during these trials were subsequently eliminated by using an improved gripper design. However, the lesson from the gripper sensor is valuable: task completion was only possible because of the operator's ability to recognize problems, react to failures, and adopt alternate strategies.

Completion times showed considerably greater variability than when the same tasks were performed in a test tank. Unlike the test tank, the real sea floor showed a range of textures (from hard rock to soft mud) and a range of gradients (slopes approached 30 degrees in some cases). Also visibility was much poorer—dropping off sharply as distances from the camera increased.

The relatively poor visual imagery available from the subsea cameras was caused by the usual underwater problems of low light levels and particulate matter in the water (this was aggravated by the necessary continual vehicle thruster action and the close proximity of lights and cameras). It was observed that the digitized images were not making full use of the available dynamic range and, as a result, software contrast enhancement was added to the slave site. Greyscale values in the range [64,255] were expanded to fill the available range [0,255], giving a visible improvement in image quality. At present the contrast enhancement is hard-coded. A simple improvement would be to use the first few images taken on-site to compute optimal contrast enhancement parameters and then use those during subsequent task execution. It is possible that further improvements could be obtained by using more computationally expensive techniques, however frame-rate considerations preclude their use given current hardware.

11.5 The November experiments

For the November experiments [71], the operator station was returned to Philadelphia while the slave site was maintained at Woods Hole, Massachusetts. Now, for the first time, the operator and remote stations were truly distant. An Internet connection was employed between the slave and master sites and an additional communications delay was inserted via software to simulate the effects of an acoustic subsea link. Image compression/decompression algorithms were now employed and the system was instrumented to enable better data collection. All signals transferred between sites were recorded and timestamped, the slave and master sites were continuously videotaped.

These trials were conducted toward the end of November. Thus, the weather conditions were, perhaps somewhat predictably, very poor. Wind speeds exceeded thirty knots and a small craft advisory was in effect for much of the time. While the vehicle was seven meters below the surface it was not immune to surface effects (especially since surface winds had a direct effect on the tether). Vertical thrusters were used at 100% continuously in an effort to maintain position and this action, coupled with rough conditions, made for very poor visibility (at times less than 1 meter).

11.5.1 *Examples of task execution*

In this subsection, results are presented from three different trials, using three different transmission time delays, to demonstrate the variability present during task performance.

An example of error-free task performance

Figure 11.9 shows an example of a successful retrieval operation. In this case, the round-trip transmission time delay was 15 seconds. The task of moving toward, grasping, and raising the capsule was encoded and transmitted to the slave as a sequence of 27 discrete commands during a 33 second period. The slave began executing those commands as they arrived. The first response (generated by the slave after completion of the first command) came back to the operator station 20 seconds after he or she first began task performance. Note that the slave and operator are working in a pipelined fashion. The operator performs the task and the slave follows along behind.

Analysis of the bandwidth requirements for this example shows that commands sent from the operator station to the slave required, on average, 0.8 kbits/second, while the replies from the slave required 0.3 kbits/second. The largest bandwidth requirement is for the transmitted image fragments. During the 35 seconds in which the slave was actively performing the task, there were 10 image fragments transmitted at an average bit rate of 6.7 kbits/second.

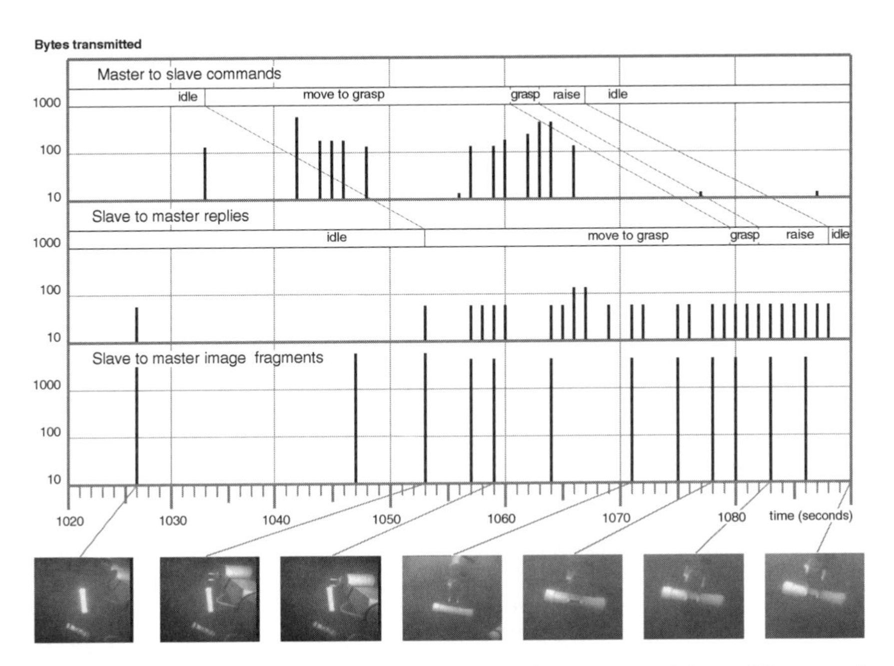

FIGURE 11.9: An example of task performance with a 15 second round-trip time delay. The graphs show the number of bytes transmitted during each second of task performance. Also shown are some examples of the image fragments automatically selected by the system for transmission back to the operator station.

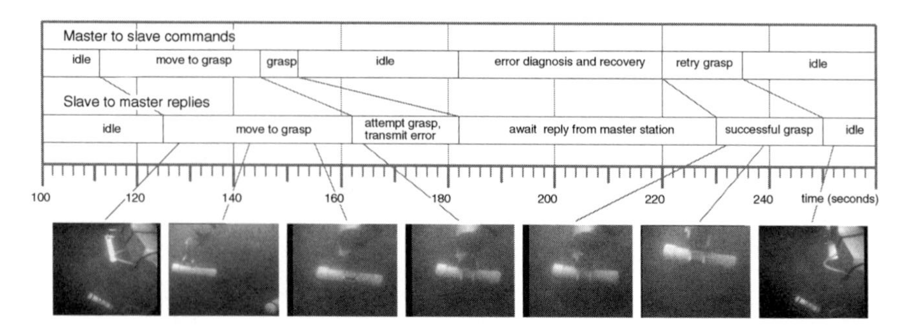

FIGURE 11.10: An example of task performance with a 10 second round-trip time delay (see text for details). Note that the depicted images represent only one third of the 24 fragments transmitted during task performance.

Simple example of error recovery

Figure 11.10 shows one of the simplest examples of task performance where an error occurred. In this case, an error was automatically detected at the slave, which paused and took new images (to assist the operator in error recovery) before sending an error message. The operator diagnosed the error "gripper torque limit exceeded" as being caused by the gripper penetrating the mud under the object while closing. He or she then raised the end effector, repeated the grasping operation, and was successful in performing the retrieval task.

Complex example of error recovery

This third example shows a situation where many errors occurred, yet the task was still ultimately successfully completed. In this case, the operator had considerable difficulty in orienting the gripper with the capsule. The angle of the object, coupled with the lack of a sixth degree of freedom, made for some awkward positioning and several attempts were required before a successful grasp was completed. During this run there were five errors, two caused by grasp failures being automatically detected, two initiated by the operator following an observed failure in the grasping action, and one caused by a torque limit as the arm contacted the vehicle during the action of raising the flask.

Figure 11.11 shows recordings of command and reply bandwidths along with the delay between the time commands were sent and the time replies were received from the remote site. Total task completion time was 17 min-

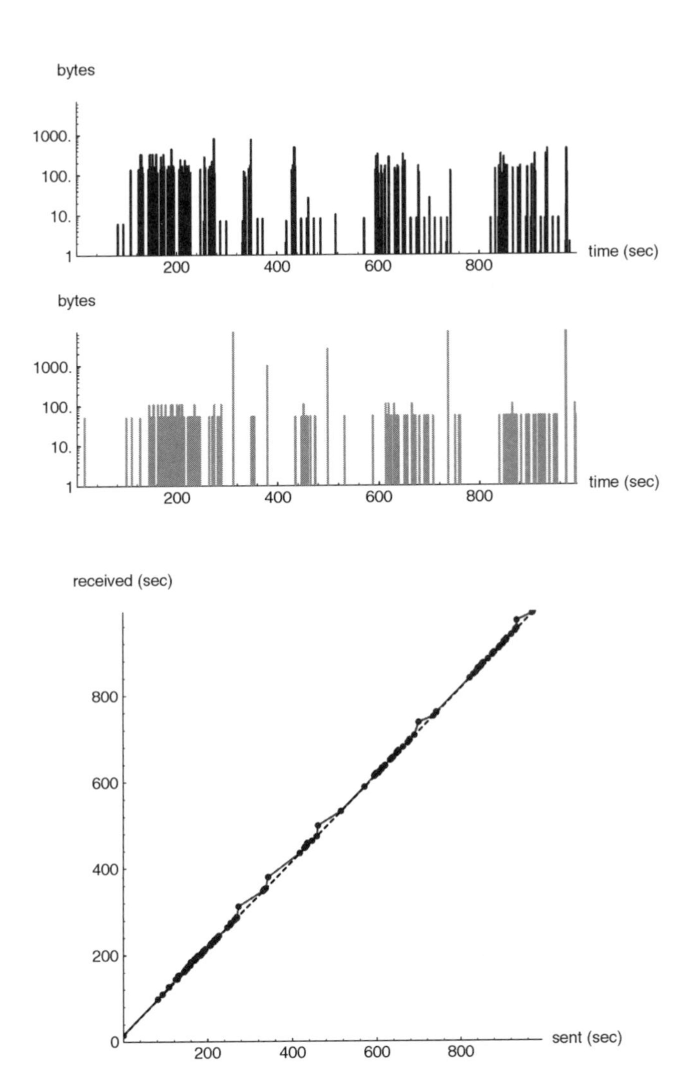

FIGURE 11.11: An example of task performance with a 5 second round-trip time delay (see text for details).

utes. Analysis of the bandwidth requirements shows that the peak bandwidth requirement was for the error messages transmitted from the slave site. This is not surprising, since these error indications contain details of past slave motion. In these tests, signal transmission was not limited by available bandwidth so these large peaks were possible; however, the system has been designed to cope with constrained channels, particularly in regard to transmitting error replies (see Section 10.2). Thus, the system could still have operated effectively with much lower bandwidths. Typical slave-to-master replies were between 50 and 100 bytes/second depending on whether one or two commands were completed per second. Peak master-to-slave transmission bandwidths were around 1 kbytes/second, corresponding to cases where several commands were sent during one second.

Comparison of times for command transmission and reply reception showed that the slave was keeping up with the operator's commanded actions. There were cases where the slave was slower. (For example, it never completed more than 2 commands/sec even though there were several cases when 3 or 4 commands were generated within 1 second.) However, it was able to eliminate the time differential during pauses in operator activity. The most significant lags are evident during commands in which errors occurred. This is to be expected, since the slave system is consumed by the task of taking new imagery of the error during this time.

11.5.2 Observations

Ideally, the speed of task performance would be limited by either the speed with which the operator could perform the task in the virtual world, the speed with which the slave could perform it in the real world, or the available communication bandwidth. In practice, given the current implementation it is the available computational hardware that is limiting performance. Faster times could certainly be achieved using more powerful machines; lower bandwidths could certainly be achieved with better compression algorithms.

In 21 trials, conducted during two days, there were only six cases where no unexpected events occurred and the capsule was successfully retrieved on the first attempt. We were clearly some distance from the safe and predictable confines of a laboratory.

There were seven cases where unexpected events occurred that prevented the task from being completed on the first attempt. These were occasionally the result of operator error (attempting to grasp the capsule with the gripper poorly oriented), but more often they were the result of unexpected vehicle and/or flask motion. In these cases, the operator was able to successfully diagnose, and recover from, each unexpected event. These represent situations where a simpler system, which did not allow for operator-mitigated error recovery, is unlikely to have been successful.

During three of the trials unexpected events occurred that were detected and diagnosed by the operator but that he or she was unable to recover from without the assistance of an operator at the remote site. These instances point to weaknesses in the specific hardware and software employed—given a slightly more sophisticated system (specifically, a system where the remote arm could be reinitialized and recalibrated without the assistance of a local operator), recovery would have been possible entirely under control of the teleprogramming operator.

In addition to those trials where the task was completed, there were five trials where the operator determined task completion to be impossible. This was the result of the object being out of sight (once), out of reach (twice) and excessive vehicle motion (twice). After the two failures due to excessive vehicle motion (the thirteenth and fourteenth trials), a flat plate was removed from the underside of the vehicle. This allowed the vehicle to settle into the sea floor, reducing motion, and allowing the experiments to continue despite worsening weather conditions.

During these trials, the poor conditions meant that the system was operating right on the boundary between the possible and impossible. As the weather conditions became a little worse, task performance became completely impossible, while as conditions improved, performance became relatively easy. In earlier experiments most problems were caused by the physical characteristics of the gripper. For these experiments, a re-designed gripper eliminated most of those problems and now, due to poor weather, vehicle motion became the major concern.

Since the system was designed with the assumption that the vehicle would be stationary, it is not surprising that a significant number of errors occurred when it was employed in an environment with significant vehicle motion. What is interesting is not so much the number of errors that occurred, but instead the number of cases where the operator was able to complete the task despite the fact that he or she was using the system under non-ideal conditions.

While there was no way for the operator to compensate for large, unpredictable vehicle motion, it was possible for the operator to detect when it was happening. This was achieved in three ways:

- Images from the remote site are taken by sequentially snapping and transmitting images from the different cameras. If the vehicle is in motion, then the pictures will have been taken from different positions and thus the relative location of the capsule in each image will be different. This becomes obvious to the operator as he or she is unable to find any position of the simulated object that is consistent with the imaged position of the capsule in all the real pictures.
- After an error occurs, the operator can overlay the latest imagery from the remote site on the virtual model. If the position of the capsule matches, then the vehicle has probably not moved. If it does not match, then either the capsule or vehicle has moved since the model was last updated.
- The operator can directly observe motion by observing image fragments transmitted from the remote site. He or she can either do this while teleoperating, or, alternatively, step through previously recorded images during error recovery.

11.6 Future implementations

In all of the experiments, it was assumed that the remote slave manipulator was mounted on a stationary platform. The obvious next step is to perform operations from a moving platform. In this case, the operator must command both the remote manipulator and the vehicle to which it is attached. Even without the effects of constrained communications, this can still be a difficult task:

> "Wilson handed him the control box of twelve toggle switches to operate *Alvin*'s arm: Grip, Wrist rotate, Elbow pivot, Shoulder rotate ... It was a real trial. Schlee worked at it for half an hour. When he twisted the robotic hand the whole submarine twisted with it. No one could do any better as long as *Alvin* hovered neutrally buoyant without a second arm to brace itself."
>
> *—Water Baby: The Story of Alvin. Victoria A. Kaharl, Oxford University Press, 1990.*

The exertion of large static forces from a hovering platform is unlikely to ever be practical; however, it should be quite feasible to perform those tasks which require more finesse than force.

In particular, it should be possible to measure vehicle motion and servo the manipulator to compensate. Using a teleprogramming style interface (with or without time delay), one could imagine a semi-autonomous system where the operator controlled the robot end-effector and the system controlled both vehicle and end-effector to perform the desired motion. This is a particularly attractive solution. It strikes the right balance between automation and manual intervention.

12

Discussion

> "The many levels in a complex computer system have the combined effect of cushioning the user, preventing him from having to think about the many lower-level goings-on... A passenger in an aeroplane does not usually want to be aware of the level of fuel in the tanks, or the wind speeds, or how many chicken dinners are to be served, or the status of the rest of the air traffic around the destination... it is when something goes *wrong*—such as his baggage not arriving—that the passenger is made aware of the confusing system of levels underneath him."
>
> *—Gödel, Escher, Bach: An eternal golden braid, D. R. Hofstadter, Penguin Books, 1980.*

A recurring question in the design of teleprogramming systems is, at what level should the human operator interact with the machine? In the present implementation, only low-level interaction is possible—the decision to use such an approach being made based on our intuition that performing tasks in the real world (and particularly recovering from errors) would be particularly difficult and unsuitable to automated approaches. It was our feeling that, if tasks could not be performed when a human was making all the high-level decisions, then they probably could not be performed at all.

A related question is, how much should the machine assist the operator? Even if the master station knew exactly what sequence of actions the operator desired, there may still be an advantage in actually having the operator perform those motions. The reason is that, if an error occurs, the operator may be better able to diagnose it after having performed the same actions themselves. In the current system we compromise. In cases where the motion sequence is known exactly (for example, when a command fixture is activated), the system simulates the motion sequence for the operator as the commands are generated. Thus, the operator is aware of exactly what actions are being commanded of the slave robot, even though he or she did not explicitly perform each one.

It is not yet clear that the expected benefit from this approach (reduced error recovery times) exceeds the cost in terms of the additional operator attention required during task performance. It may be faster to attempt automated solutions first and then acquaint the operator with what the system tried and what happened only in those cases where failures are evident. This need to reacquaint the human with what the system was doing will become more important as the efficiency of the operator interface is increased. As the operator is able to perform tasks faster, there will be a natural progression toward having a single operator control more and more robotic devices. As the amount of time spent interacting with any particular robot during any particular task decreases, so the need to aid the operator in diagnosing problems with that task will grow.

A related question is, how complex should the slave system be? In this case the answer is easier. The presence of an intelligent operator within the system means that there will nearly always be an advantage in spending limited resources on sensors rather than remote processing power. Provided the operator has sufficient information to make a correct diagnosis, then we can rely on his or her reasoning ability to compensate for any inadequacies in processing at the slave site.

12.1 Bandwidth considerations

The present implementation was designed with the goal of operating with communications bandwidths on the order of 10 kbits/s. Should the teleprogramming paradigm be applied to other domains, it is quite likely that larger bandwidths will be available. (In fact, even in the difficult case of underwater communication, much greater bandwidths may one day be possible [81]).

A much larger bandwidth would make it feasible to consider the use of color cameras (or other sensors) with faster update rates and larger imaging regions. It would also make it feasible to consider taking more than one image (or other sensor) reading at each instant and sending multiple parallel data streams to the operator station. While it may be unwise to display all that information to the operator, it would be useful to have additional information for those cases where errors occur during execution.

In the current system, the operator station could attempt to predict those cameras that would have had a good view of the error event, but that information is not useful because the event has already occurred and it is too late to take additional images. However, if a larger bandwidth were available, the slave could send more than one image fragment for each instant in time. In this case, after an error message had been received, the master station could select from among several different alternative images when displaying the event for the operator. Here, there is considerably more flexibility since the temporal constraint imposed by the need to keep up with the continuous motion of the slave is removed.

This process—of choosing from among multiple images of the same event—is known in cinematographic terms as Complexity Editing [99]. Two techniques are especially relevant: the first, known as a Sequential Analytical Montage, involves showing different scenes in sequential order (although not necessarily with the same time scale) so that cause-and-effect relationships become apparent. The second, known as a Sectional Analytical Montage, is where several different views of the same period of time are presented. Here the intention is to give the viewer more information about a single event than could be provided from just a single viewpoint.

Just as a larger bandwidth would allow multiple images to be sent at the same time, so it would also allow multiple commands to be transmitted. In the present system the limited bandwidth imposes a significant cost to each transmitted command. Thus, it is unwise to send commands to the slave unless it is likely that they will be executed. However, if a much larger bandwidth were available, then that could be used to mitigate, to a certain extent, the time delay. The idea is to send a tree of commands to the remote vehicle. Based on its sensory readings, it chooses one path through that tree of commands for execution. The operator station, by observing the received replies, could recognize which branch had been taken and transmit subsequent commands starting from the end of that path. Thus, the slave has the opportunity to react to sensory readings, choosing different execution streams even though the operator station won't find out which stream until at least one communications time delay later. This strategy depends on the operator station translating operator action into not one, but several command streams based not only on the expected success of operations, but also on likely failures.

12.2 Programming by demonstration

In the present incarnation, the operator station has a memory of past operator actions but it uses that only for displaying action sequences to the operator during error recovery procedures. This is adequate while tasks are short and non-repetitive. But, as tasks become more complicated and are repeated multiple times, it will be necessary to consider more sophisticated approaches.

Current teleoperation systems allow operators to record and replay motions [14] (much like programming macros). A clue to possible improvements comes from the literature on demonstrational interfaces [56]. In such interfaces, the system attempts to construct a parameterized program from observations of operator interaction. Cypher's "Eager" system [19] for the Macintosh Hypercard program, is particularly elegant. It looks for repetitive operator actions and, once a cycle is detected, it signals its knowledge to the operator by predicting what he or she will do next (by, for example, highlighting the menu item that the operator is expected to select). Once the operator, by repeating the task and watching the predicted symbols, is confident of the system's interpretation, then he or she may relinquish control to the system, allowing it to continue the process automatically.

For teleoperation, the ultimate need to control a real manipulator means that the consequences for an incorrect interpretation of operator action are much more severe than in the case of a pure software environment such as Hypercard. Nevertheless, by providing graphical simulations of programmed motions [97, 18], it should be feasible to allow operators to check parameterized sequences prior to execution. The system could, for example, display a semi-transparent copy of the manipulator at the predicted position a few seconds ahead of the currently commanded action. So long as the operator was satisfied that the system's predicted motions were acceptable, he or she could allow it to continue autonomously.

12.3 Learning experience

Working with robotic systems to perform even very simple tasks gives one a deep appreciation for the complexity of real-world interaction. It will be some time before autonomous systems are able to replace human operators for performing manipulative tasks in unconstrained real world environments.

> "It is very hard for any adult to remember or appreciate how complex are the properties of ordinary physical things... To catalog only enough knowledge to enable a robot to build a simple blocklike house from an unspecified variety of available materials would be an encyclopedic task. College students usually learn calculus in half a year, but it takes ten times longer for children to master their building toys. We all forget how hard it was to learn such things when we were young."
>
> *—Marvin Minsky in The Age of Intelligent Machines, Raymond Kurzweil (ed), M.I.T. Press, 1990.*

12.4 Interacting with uncertainty

Throughout this text, there has been the recurring problem of how to cope with the uncertainty associated with real-world interaction. In Chapter 7, it was argued that fixtures could reduce both the system's uncertainty as to which command the operator desired and the operator's uncertainty as to which command the system would generate. It was further argued that the filtering function of synthetic fixtures could prevent positional uncertainty at the master arm from flowing through to the remainder of the system.

The visual imaging system, developed in Chapter 8, reduced the operator's uncertainty as to the state of the remote site by providing appropriate image fragments to assist in error diagnosis and recovery.

In Chapter 10, the uncertainty associated with the specification of motions at the remote site was discussed and it was noted that specifying motions as desired positions relative to symbolic locations could reduce uncertainty by preventing the positional errors in each motion from accumulating.

The mechanisms used to mitigate the effects of uncertainty have been *ad hoc* approaches developed through common sense and intuition. While this was appropriate for the present implementation, there is much room for future work (particularly in evaluating characterizing distributions).

12.5 The virtual reality mirage

In simulating the remote environment for the operator it is tempting to desire more and more realism. But reality is a hard thing to copy. At least for now, it does not seem likely that we will ever develop a system so perfect that an adversarial user might not be able to distinguish it from reality. Perhaps more importantly, it is not clear if we should even be attempting to provide realism.

In Chapter 3, it was noted that simplified graphics displays offered superior performance. In Chapter 7, it was argued that unrealistic force and visual clues could offer improved performance. Thus, at least for teleoperation applications, the provision of very realistic feedback does not appear a desirable goal.

12.6 Future interfaces

Brooks et al. concluded that smaller devices, feeding forces to fingers-hand, rather than hand-arm, were preferable because they were less tiring, of at least as good resolution, were simpler, economical, and smaller [11]. Experience with the teleprogramming system also supports this view.

In present implementations, the operator must assume much of the responsibility for planning motions of the slave robot. Thus, there is an advantage in using input devices that allow him or her control over all the remote degrees of freedom. This will change. As our systems become more intelligent, the operator will be able to interact in increasingly more abstract terms. It will no longer be necessary for the operator to actually perform desired motions. Instead, he or she need only provide enough input to allow the system to infer their intentions. It is expected that input devices with only two or three degrees of freedom will become common.

12.7 The distant future

It's tempting to believe that as our computers keep getting faster, so our machines will keep getting smarter. This is not necessarily true. For most of the fundamental problems remaining in robotics, it's not that we don't have enough computational power; rather, it's that we don't know what to compute. There is much work yet to be done.

A five-year-old child can survive in an unfamiliar environment, while our most sophisticated machines consistently fail at the first opportunity. Some would argue that this is because our computers are lacking in power when compared to a human brain. Even if this is true now, it will not remain so for long. Our computers are growing in power exponentially while our brains are standing still. Its only a matter of time before the former surpasses the latter.

It is quite likely that, within ten years, we'll have machines with more storage capability and more raw processing power than a human brain. The challenge now is not to build bigger, faster, machines. Instead it is to figure out what to do with all that processing power. I don't want a computer that can balance my checkbook a million times faster than I can. I want a robot that can find my checkbook.

13

Conclusions

The teleprogramming paradigm allows an operator to control a remote robot efficiently despite significant communication delays. The operator interacts with a virtual reality representation of the remote site that provides immediate visual and kinesthetic feedback. By monitoring the operator's actions, the master system is able to automatically generate a symbolic, error-tolerant, command stream for transmission to, and execution by, the remote manipulator.

Synthetic fixtures have had a pervasive effect on the system. By providing force and visual clues, they assist the operator in performing actions with speed and precision. By helping the operator to make deliberate, non-ambiguous, motions they simplify the task of translating action into robot commands. By guiding the operator's choice of actions, they help him or her avoid the generation of commands that are likely to fail in the presence of positional uncertainty. A number of fixtures have been implemented. Examples have been presented showing both simple single-fixture applications as well as complex multi-fixture systems.

By keeping commands simple (a single action) and short (around 500 ms), we simplify the slave implementation, while making it easier for the operator to determine causality should an error occur. Complex actions are performed by transmitting a sequence of simple commands. This gives the master station some control over how complex actions should be performed and allows the operator to select an appropriate level of command conservativeness—choosing between slow and safe or fast and risky command sequences.

The use of real visual images improves the operator's situational awareness and gives him or her a powerful tool for diagnosing errors. Since the available communication bandwidth is limited, the system must employ an intelligent approach. It automatically determines which fragment of the available imagery can best aid the operator at each instant. It dynamically varies the frame rate depending on the task actions being performed, and it alters the rate of task execution to match that which the imaging system can support, given available bandwidth and computational resources.

We have strived to develop a system that employs technology appropriately. It automates tedious and time-consuming operations while relying on a human operator to provide the reasoning and intelligence necessary to cope with the wide variety of unexpected events that occur in real-world interaction.

Working systems have been implemented in air, in a test tank, and in the ocean. These tests have demonstrated that real-world intervention can be performed in unpredictable environments even while using delayed low-bandwidth communications links.

In the near future, we can expect significant improvements in remote control systems that operate via the constrained communications of the Internet, particularly those using World Wide Web style interfaces. These will reimplement the ideas from traditional teleoperation (and perhaps even a few ideas from this book). Expect visual clues, predictive displays, and automated intelligent control of remote cameras.

Teleprogramming has raised the level of communication between master and slave stations. In the future, the level of communication between operator and master system will also become more abstract. As this happens, it will no longer be necessary for the operator to deliberately perform every motion. Complex hand controllers (such as the small Puma 260 robot which our operator held), will become unnecessary, and the remote control of multiple manipulators from a single operator station will become commonplace.

In the not too distant future, our robots will no longer be limited by available computational power. The big new advances in remote control robotics will not come from having faster computers, they will come from the invention of better algorithms.

Appendix A

Implementation

This appendix describes the implementation of the teleprogramming system used for the subsea trials. It is this implementation that coalesces the operator interface, synthetic fixtures, command generation, visual imagery and error recovery into a working system.

These details are included in the text for two reasons. First, they add some concreteness to the discussion, moving us well away from abstract unrealizable ideas. Second, they provide a means to pass on our experiences. It is hoped that those who follow us may avoid at least a few of our more obvious mistakes.

We begin by looking at the operator interface and the way in which the operator's actions within the virtual world are transformed into commands. Then we consider how those commands are transferred to the remote site and how they are interpreted and executed. Finally, we examine the mechanism for returning symbolic replies to the operator station and the procedures for error recovery.

A.1 Operator interaction with the master station

The teleprogramming master station combines a graphics workstation with a commercial robot manipulator to provide a system capable of providing real-time visual and kinesthetic feedback to a human operator (see Figure A.1).

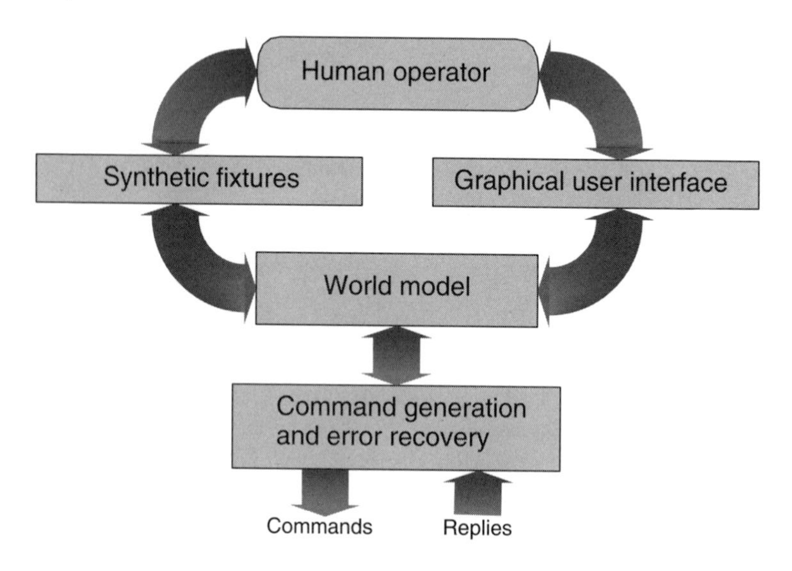

FIGURE A.1. Logical overview of the master station.

The operator commands the system by moving a control ball mounted at the end of a Puma 260 robot arm. A 6-axis force/torque sensor mounted in the control ball senses the operator's motions and the robot arm is servoed accordingly. Kinesthetic feedback to the operator is provided by actively servoing the arm to resist the operator's commanded motion. Low-level arm control hardware is housed in an IBM-PC-AT. This machine communicates with a JIFFE high-speed processor [4] which performs all robot control computations. The JIFFE, in turn, communicates with the Silicon Graphics workstation, which assumes responsibility for high-level control while providing visual feedback to the operator.

The graphics machine communicates with the JIFFE coprocessor via a BIT-3 VME-VME interface. This coprocessor provides 40 real Mflops of computational power [4]. It is used in the teleprogramming system to run the control software for the robot arm, implementing a PD controller with feedforward gravity compensation at 500 Hz while running a Cartesian servo loop and performing all necessary fixturing calculations at 30 Hz.

The JIFFE communicates with the PC-based robot controller using shared memory via a second BIT-3 interface. It is connected to a modified Puma 560 controller box via an interconnection board incorporating a custom-designed hardware watchdog timer. Two TE5312 Encoder cards (4 axes each) are used to record position for the six robot joints, while a DDA-06 DAC and digital I/O board is used to provide analogue outputs for each joint motor as well as for testing and configuring the safety circuitry. A parallel/serial I/O card provides a serial port that is used for connecting a LORD LTS-200 force/torque sensor (the sensor is mounted in the control ball at the end of the robot arm). Processing power for the controller comes from an old 8 MHz IBM-AT motherboard. The processing consists of simply copying data to/from dual-port memory and the encoder, DAC, and serial interface cards while continually checking the status of all components.

While the implementation of the JIFFE software is complicated by its single-threaded nature, the underlying algorithms are relatively simple (the equations for remapping motion of the master arm based on the operator's chosen viewing direction and generating force clues for active fixtures were presented in Chapters 6 and 7). It is within the graphics workstation that the most interesting software structures appear, and it is these structures that will be the focus of the remainder of this section.

A.1.1 The user interface

The user interface is implemented entirely as a push down automaton [37]. Each state of this machine describes a particular display configuration (buttons, windows, dialog boxes). Transitions between states may be initiated by the system (such as when an error is detected at the remote site during teleprogramming) or by the user (such as when he or she presses an on-screen button). The mechanism for displaying and managing the user interface components is particularly simple. A text file describes the system states and allowable state transitions. When in a particular state, the system examines allowable transitions to determine which options to make available to the user. For the teleprogramming implementation, this interface contains approximately 50 states and 100 transitions.

A.1.2 The world model

The world model is represented by two distinct hierarchies. The first, an object-oriented class hierarchy, records an *is-a* relationship [17]. Hard-coded within the system software, it describes interrelations between the different classes of shapes. The second, a model hierarchy, records a *de-*

pends-on-the-position-of relationship. This is dynamic and changes during task execution as objects are moved within the virtual world.

The class hierarchy

The code to implement the shapes from which the virtual world is constructed is written in C++ [82]. This software encodes the class hierarchy. For example, in the current system, the world model is composed of `ptShapes`; the subset of `ptShapes` that have vertices are termed `ptObjectShapes` and the subset of those whose vertices lie on a plane are termed `ptPolygons`. Thus, every `ptPolygon` **is-a** `ptObjectShape`, and every `ptObjectShape` **is-a** `ptShape`. This hierarchical structure allows for efficient coding since the software to perform an operation on any subset need only be written once for the parent of that subset. Figure A.2 shows examples from the class hierarchy used in the teleprogramming implementation.

The shapes required for teleprogramming fall naturally into six categories.

`ptRootShape` This shape is unique. In each instantiation of the world model, there is only a single root shape.

`ptSimpleShape` These are those shapes that are added to the world model (usually temporarily) to provide information to the operator. For example, when the operator moves near a joint limit, a warning marker is added to the world model. That marker **is-a** `ptSimpleShape`.

`ptLinkShape` These shapes represent dependencies within the world model. They take the form of fixed transforms or variable joints and change the position, orientation, or scale of their children within the world model hierarchy.

`ptObjectShape` These represent shapes that model objects in the real world. Typical examples are cones, cylinders, cubes, and polygons. Many of these shapes are stored in a single fixed size and preceded by `ptLinkShapes` to scale them to the desired shape.

`ptMaterial` This shape is for purely aesthetic considerations. It changes the material properties of rendered shapes. Note that, for implementational reasons, the `ptMaterial` affects every subsequently rendered object—not just those which are beneath it in the world model hierarchy. This departure from a strict hierarchy is desirable here because, unlike transformation matrices, material properties cannot be efficiently pushed and popped during recursive descent traversal.

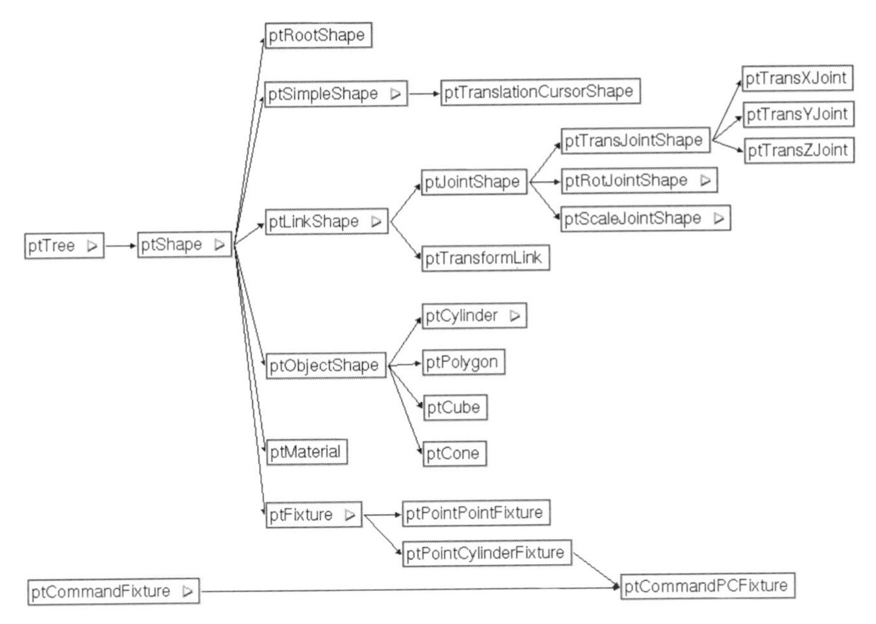

FIGURE A.2: The Shapes class tree. Note that only a few representative examples are shown—the ▷ indicates cases where some derived classes are not shown.

`ptFixture` These shapes code descriptions of different synthetic fixtures. They differ from other shapes in that they determine their own visibility within the world model and may optionally alter the positions of other shapes within the world model.

Note that the hierarchy is not a pure tree since a node may have more than one parent. This case is termed *multiple-inheritance* [82]. It occurs here because command fixtures serve two roles. First, they act as regular fixtures, providing force and visual clues to operators, and second, they activate functions upon being penetrated by the end-effector.

The world model hierarchy

The world model is constructed as a hierarchical tree of shapes. More formally, it is a directed acyclic graph [2] where each node (except a unique "root" node) has exactly one parent node, any number of child nodes, and a unique path back to the "root" node. The model may be rendered by recursively descending through the tree, having each node draw itself and then invoking the draw procedure for each of its children.

Each shape within the tree has two associated transforms. The first, represents the absolute translation, rotation, and scale, in world coordinates for vertices within this shape. The second, relative transform, represents the differential change between the absolute transform of its parent node and its own absolute transform.

The location within the virtual world of any shape (equivalently, the absolute transform of any node within the world model tree) depends only on those shapes along the unique path between that node and the root node. This property allows the implementation of an efficient lazy evaluation scheme for computing the location of any vertex. When any shape is modified in a way that may affect the locations of its children (for example, if a joint shape is rotated), then all of the descendants of that shape are marked as having an incorrect position.[1] Then, if the position of any shape is required, the system works back up the tree from that shape until it finds a node whose position is known to be correct. It then works back down the unique path to the desired shape, evaluating transforms for shapes along the way to compute the desired position.

In addition to the tree representation, the system also maintains a series of parallel representations in the form of doubly-linked lists—one list for each logical subset of shapes. These lists serve a dual purpose. They allow some operations to be more efficient and also make the implementation more "pure" in the sense that it removes much of the need for type casting. For example, it is often necessary to find the locations of all vertices within the world model. There are three ways to do this:

1. Create a virtual function `getVertices()` that is defined for each `ptShape`. Have it return an empty list by default, but for `ptObjectShapes` have it compute and return a list of all vertices. Then, to find all the vertices in the world model, we can recursively descend through the world model tree, calling `getVertices()` for every node. This will work, but it is undesirable because the implementation of the generic `ptShape` class is now dependent on the specific features needed for `ptObjectShapes`. Every time a new feature is added it will again be necessary to modify the definition for the generic shape.

2. Another solution is to define a function `getVertices()` for the `ptObjectShapes` class. Then the system can recursively descend through

[1]This recursive descent marking procedure is efficient since it need not descend past any node which is already marked as being incorrect.

the world model tree. For each node, check to see if it is a `ptObject-Shape`. If so, then cast that node to be a `ptObjectShape` and call `getVertices()`. This avoids the need to change the definition of `pt-Shape`, but it introduces the need to check types and cast pointers—both of which are common sources of programmer error.

3. The third possibility, and the one currently employed, is to maintain a list of all `ptObjectShapes`. Then, when a list of vertices is required, the system can simply step through that list, calling `getVertices()` on each element. This implementation is more efficient since there is no need to recursively descend through every shape. It is also more "pure", removing any need to check types or cast pointers. There is some overhead in maintaining these lists. However, since the number of objects in the world model does not often change, this overhead is relatively small. The lists are maintained automatically. Whenever a new shape is constructed, it is added to the appropriate lists and similarly, when a shape is deleted, it is removed from all relevant lists.

A.1.3 Interpreting operator action

As the operator controls the simulated slave robot, his or her motions are monitored and translated into a sequence of robot actions. Stored in a doubly-linked list, these actions encode the system's interpretation of operator motion. As each action is stored, it has the option of generating a command sequence for transmission to the real remote robot. The class tree used to implement the possible actions is shown in Figure A.3.

In the present implementation, there are eight basic types of actions:

`ptRobotNullAction` This indicates that the operator has performed no action during the preceding few seconds of task performance. It causes generation of an empty command for the remote site. While these generated commands do not explicitly request any action, they implicitly require the slave manipulator to remain stationary. They also serve as markers in the command stream, ensuring that there is an upper bound on the time between successive command transmissions.

`ptRobotMotionAction` This is the most commonly generated action. It encodes a Cartesian motion of the end-effector. These motions are generated every time the system checks the end-effector position and observes that the operator has moved it. Thus, the resolution to which

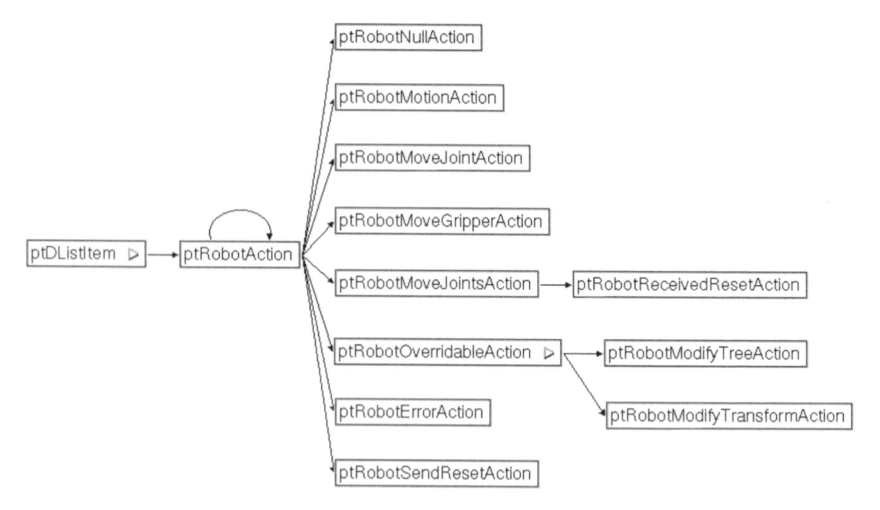

FIGURE A.3: The robot action class tree. The loop-back over the `ptRobotAction` indicates a self-referential definition—each `ptRobotAction` contains pointers to past and future `ptRobotActions`.

these discrete actions represent continuous operator motion is limited only by the processing speed of the machine employed. However, as previously mentioned, sending hundreds of very small motion commands to the slave site is inefficient. So, the system looks at the differential change (both spatially and temporally) between the last transmitted motion command and the current motion action to decide whether the transmission of a command for the current action can be justified. Thus, the conversion of actions into transmitted commands is a many-to-one process. Several motion actions may result in only a single generated command.

Some care is necessary in determining exactly which motions should cause new commands to be sent to the remote site. For example, consider the action of moving into contact with a surface at the remote site whose position/orientation is not precisely known. Assume that, with high probability, the surface is known to lie within a fixed bounding region. In this case, it would be inadvisable to command a motion that ends at a point inside the bounding box but which does not contain an expectation of contact with the surface. Such motions may be avoided by using the filtering function of synthetic fixtures.

`ptRobotMoveJointAction` This action is created whenever the operator chooses to move a single joint on the simulated slave. The generated command instructs the real slave to move the same joint by the same amount.

`ptRobotMoveGripperAction` Whenever the operator opens or closes the gripper this action is generated. In this case the generated command for the remote manipulator depends on whether the gripper is being opened or closed and the size of any object on which it may be closing.

`ptRobotMoveJointsAction` This is created when the operator performs a joint space motion with the remote manipulator. The generated commands instruct the slave robot to assume the same joint space position.

`ptRobotOverridableAction` This class of actions encodes changes to the world model as a result of operator action. For example, when an object is picked up, the structure of the model changes. The position of that object now depends on the position of the end-effector, rather than the position of the piece of model on which it previously rested. These actions do not cause the generation of any commands for the remote slave robot. Instead, they serve as markers in the robot action stream so the system may "remember" that point in time when changes in the position or structure of objects occurred within the model. This memory is important since, without it, the system would be unable to correctly "wind back time" in the event that an error was detected. These actions are overridable since the operator will have the opportunity to examine the system's interpretation of the state of the real remote site and, if necessary, correct any invalid conclusions.

`ptRobotErrorAction` This action is generated whenever the operator forces the system to initiate error recovery procedures. This causes generation of a command for the slave robot that forces it to pause and enter an error recovery state (just as though it had detected an error itself). This is necessary to cope with the case where the operator notices an error that the slave failed to detect.

`ptRobotSendResetAction` This action is generated as the operator begins to command the robot after recovering from an error. It causes generation of a reset command for the slave robot; indicating that subsequent commands were generated after the operator was aware that an error occurred.

A.2 Master-to-slave communication

The master (operator) station communicates with the slave (subsea) site by transmitting a command stream composed of execution environments. The full definition for this command stream is given in Table A.1.

This section will begin by describing the way in which joints, sensors, transformation matrices, and frames are defined. It will then describe the command stream, showing how motions are specified, how sensors are interrogated, and how frames are defined.

A.2.1 Basic definitions

Cameras

camera → **c***integer*

Each camera is uniquely numbered. At present there are only two cameras c0 and c1.

Joints

joint → **j***integer*

Each joint is uniquely numbered, starting at the robot kinematic base and working out toward the end-effector. The first joint is j0, the next j1 and so on. Tools are typically treated as extra joints on the robot. For example, on the JASON arm the gripper is j5.

Sensors

sensor → **s***integer*

Each sensor is distinguished by a unique identifying integer. Sensors with identifying numbers less than 1000 are "fixed"—they are always observed by the system. If any fixed sensor becomes active, that is treated as an error condition by the slave robot. Sensors with values greater than 999 are considered "optional." They are not observed unless they are specifically mentioned within an execution environment. Predefined sensors for the JASON manipulator are listed in table A.2.

commandStream	→	*environment commandStream*
environment	→	**R** *nl*
	|	**B** *integer nl*
		preMotions
		motions
		postMotions
		E *nl*
preMotions	→	*visual preMotions*
	|	ϵ
visual	→	*camera float float integer float nl*
motions	→	*motion motions*
	|	*motions*
motion	→	**M** *joint* = *float nl*
	|	**M ee** = *position nl*
	|	**M d** = *float nl*
postMotions	→	*postMotion postMotions*
	|	ϵ
postMotion	→	**I** *sensor* = *integer integer nl*
	|	**D** *frame* = *position nl*
position	→	*frame*
	|	*frame*.*transform*
frame	→	**ee**
	|	**f***integer*
transform	→	**t(** *float float float float float float*
		float float float float float float **)**
sensor	→	**s***integer*
joint	→	**j***integer*
camera	→	**c***integer*

TABLE A.1. Language definition for master-to-slave communication.

Sensor	Description
s0	Commanded motion completed
s1	Joint 0 minimum limit
s2	Joint 1 minimum limit
s3	Joint 2 minimum limit
s4	Joint 3 minimum limit
s5	Joint 4 minimum limit
s6	Gripper minimum limit
s11	Joint 0 maximum limit
s12	Joint 1 maximum limit
s13	Joint 2 maximum limit
s14	Joint 3 maximum limit
s15	Joint 4 maximum limit
s16	Gripper maximum limit
s21	Joint 0 torque limit
s22	Joint 1 torque limit
s23	Joint 2 torque limit
s24	Joint 3 torque limit
s25	Joint 4 torque limit
s26	Gripper torque limit
s1000	Gripper contact

TABLE A.2. Predefined sensors used for JASON manipulator.

Transformation Matrices

transform → **t**(*float float float float float float float float* ... *float*)

Where the order of the floating point parameters is: nx, ny, nz, ox, oy, oz, ax, ay, az, px, py, and pz. These specify the matrix:

$$\begin{bmatrix} nx & ox & ax & px \\ ny & oy & ay & py \\ nz & oz & ax & pz \\ 0 & 0 & 0 & 1 \end{bmatrix}$$

Frames

frame → **ee**
 | **f***integer*

where ee is a special frame referring to the location of the end-effector at the end of motion for the current execution environment. All other frames act as symbolic labels for fixed locations within the workspace. Each has a unique identifying number that is assigned when it is created. The only predefined frame is f0, the kinematic base frame for the manipulator.

A.2.2 The command stream

commandStream → *environment commandStream*

The command stream is composed of a sequence of environments, where each environment either signals a reset or describes an action to be executed.

Reset environment

environment → **R** *nl*

The reset (R) environment is used to synchronize the master and slave sites after an error. It indicates that the following environment is the first of a new sequence of commands. When the slave detects an error condition, it will notify the master station and then ignore all received environments until a reset environment arrives. Once that has been received, the slave may process and execute any following commands as normal.

Execution environment

$$
\begin{array}{lcl}
\textit{environment} & \rightarrow & \textbf{B}\ \textit{integer nl} \\
 & & \textit{preMotions} \\
 & & \textit{motions} \\
 & & \textit{postMotions} \\
 & & \textbf{E}\ \textit{nl}
\end{array}
$$

Execution environments start with a begin (B) command that includes an integer command identifier. Command identifiers are uniquely assigned to each executed environment. Think of command identifiers as markers in the command stream. They allow the master and slave station to refer unambiguously to particular actions.

Each environment describes a single, simple and short action. A typical action is, in colloquial terms, "close the gripper and expect to grasp an object." On average, each action will take around 500 ms to execute.

The end (E) command signifies the end of an execution environment. Typical slave implementations will read characters from the command stream until an E is encountered, then, knowing that a whole environment has been read, they may begin to execute the described action. This procedure guarantees that the slave knows what it should do at the end of an action before it begins to perform that action.

A.2.3 Pre-motion commands

$$
\begin{array}{lcl}
\textit{preMotions} & \rightarrow & \textit{visual preMotions} \\
 & | & \epsilon \\
 & & \\
\textit{visual} & \rightarrow & \textit{camera float float integer float nl}
\end{array}
$$

At present, the only pre-motion commands are camera specifications. Each camera specification lists an x coordinate, a y coordinate, an integer scale factor, and a floating-point score. The x and y coordinates indicate the center of the region of the image that is most appropriate to the current execution environment. The scale factor is a software version of a camera zoom. A scale of one indicates a one-to-one mapping of framegrabber pixels to transmitted image pixels. A scale of two indicates that only every second pixel is transmitted, and a scale of three that only every third pixel is transmitted. Each camera is also assigned a score based on the relative importance of that view of the scene. The slave may make use of this information to vary the relative frequency with which images from each camera are taken.

A.2.4 Motion commands

$$\begin{array}{lcl} \textit{motions} & \rightarrow & \textit{motion motions} \\ & | & \epsilon \end{array}$$

Joint motions

$$\textit{motion} \rightarrow \mathbf{M}\ \textit{joint} = \textit{float nl}$$

Each joint motion command instructs the slave to move the specified joint so that it has the specified value. All joint values are specified in radians for rotational joints and in millimeters for prismatic joints. Multiple motion commands may be combined into a single environment to allow coordinated joint space motions. In such cases, motion of all joints should occur in parallel, with all joints starting and stopping in unison. For example, the command to move the JASON arm to the zero position is:

```
B 1
M j0=0
M j1=0
M j2=0
M j3=0
M j4=0
E
```

Cartesian motions

$$\begin{array}{lcl} \textit{motion} & \rightarrow & \mathbf{M\ ee} = \textit{frame nl} \\ & | & \mathbf{M\ ee} = \textit{frame}\mathbf{.}\textit{transform nl} \end{array}$$

The position is calculated by postmultiplying the transformation matrix for the specified frame by the specified constant transform. For example, to move the end-effector to a position 400 mm in z from the kinematic base frame use:

```
M ee=f0.t(1,0,0,0,1,0,0,0,1,0,0,400)
```

At present, the right-hand-side of the motion expression is very limited, but in the future much more complicated expressions should be possible.

Motion duration

$$motion \rightarrow \mathbf{M\ d} = \mathit{float\ nl}$$

The desired motion duration is specified in seconds as a floating-point number. If a duration is specified without any other motion commands, then the slave will pause, remaining stationary for the specified time. If a duration is specified with motion commands then the slave will attempt to perform the requested motions in the desired time. However, it is not guaranteed that it will do so. The slave may perform the action at a slower rate in order to avoid exceeding velocity or torque limits. If no duration is specified, then the slave will choose a "natural" speed for the desired motion.

A.2.5 Post-motion commands

$$\begin{aligned} postMotions &\rightarrow postMotion\ postMotions \\ &\mid \epsilon \end{aligned}$$

Post-motion commands are optional. They may be used to specify actions to be performed after any commanded motion has terminated.

If command

$$postMotion \rightarrow \mathbf{I}\ sensor = integer\ integer\ nl$$

At present, all sensors are binary so only equivalence relations are supported. However, in the future floating-point sensor values and full relational operators will be supported. Each If command has two effects. First, during any commanded motion the specified sensor should be observed and any change in that sensor's value should cause all motion to stop. Second, after any commanded motion has stopped and iff the specified sensor has a value equal to the specified integer value, then the system should perform an action specified by integer. If integer is 0, it should generate an error, if integer is positive, it should skip forward by that number of commands (ignoring any following post-motion commands for the current environment). For example, the following command instructs JASON to close the gripper. If the gripper contact sensor is not active, it should generate an error. If it is active, then it should continue with the next command.

```
B 1
M j5=0
I s1000=0 0
```

```
I s1000=1 1
E
```

While fixed sensors are observed for every command, optional sensors are only observed if they are specifically mentioned in an If statement. Each command thus has a number of implied If statements. For example, the preceding environment would be executed as though it were:

```
B 1
M j5=0
I s1=0 0
I s2=0 0
: :  :
I s999=0 0
I s1000=0 0
I s1000=1 1
E
```

Define command

$$postMotion \quad \rightarrow \quad \mathbf{D}\ frame = \quad position\ nl$$

This defines a specified frame to be at the specified position. Note that this definition occurs after motion for the environment has terminated. For example, the following command moves the robot to the zero position and then defines that location as frame f1.

```
B 2
M j0=0
M j1=0
M j2=0
M j3=0
M j4=0
D f1=ee
E
```

Note that defines are much more than just syntactical sugar. Since they evaluate expressions using data read from the slave, they enable the master station to program in actions that depend on sensory data just recorded at the slave site.

A.2.6 Example command stream

The following example shows the first seven environments from a JASON command stream. The slave is commanded to move to a shoulder-down configuration, then perform a Cartesian motion in the +z direction, and then open and close the gripper.

```
B 0
M j0=0
M j1=0.523599
M j2=1.5708
M j3=0
M j4=-0.523599
E
B 1
M ee=f0.t(0.0 0.0 1.0  0.0 -1.0 0.0  1.0 0.0 0.0
          933.8 0.0 -149.4)
E
B 2
M ee=f0.t(0.0 0.0 1.0  0.0 -1.0 0.0  1.0 0.0 0.0
          933.8 0.0 -81.1)
E
B 3
M ee=f0.t(0.0 0.0 1.0  0.0 -1.0 0.0  1.0 0.0 0.0
          933.8 0.0 26.3)
E
B 4
M ee=f0.t(0.0 0.0 1.0  0.0 -1.0 0.0  1.0 0.0 0.0
          933.8 0.0 82.6)
E
B 5
M j5=1.5708
E
B 6
M j5=0
I s1000=1 1
I s1000=0 0
E
```

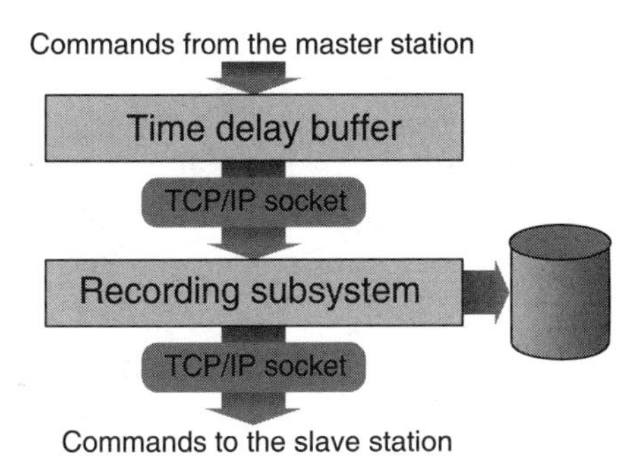

FIGURE A.4: Overview of the communication system between master and slave.

A.2.7 Telemetry

The teleprogramming commands generated by the operator station are transmitted to the slave site via a TCP/IP socket connection [80] with an additional delay inserted via software to simulate an acoustic link. A separate process is inserted into the communication channel to timestamp and record all transmitted telemetry for later use in analyzing bandwidth requirements. Thus, even though a real acoustic link is not being employed, the time delay is comparable to a real link and the communication bandwidth may be monitored for consistency with acoustic parameters. The system is shown in Figure A.4.

A.3 Command execution at the slave site

The teleprogramming slave system comprises an interpreter, manipulation, and imaging subsystems (see Figure A.5).

The interpreter serves as an interface between the teleprogramming telemetry and the manipulation and imaging subsystems. It converts the symbolic teleprogramming commands into low-level absolute joint space motions for the slave manipulator. It observes the low-level replies from those systems, recording data for transmission to the operator station, and it compares predicted and recorded sensory signals, looking for discrepancies between the expected and actual data.

For each received command, the interpreter performs the following functions:

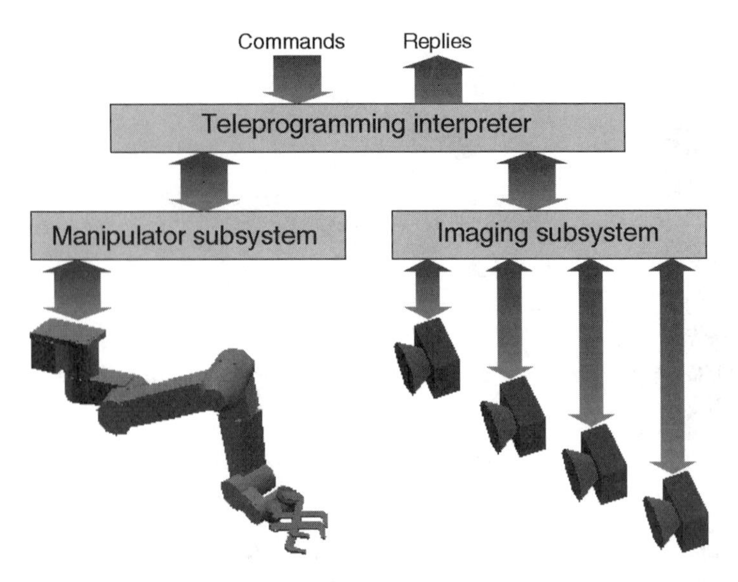

FIGURE A.5: Overview of the teleprogramming slave implementation.

1. Set the current command identifier. Any error that occurs from this point in time will be reported as occurring in this command.

2. Phase 1 parser

   - Read and perform any pre-motion commands. Since the only current commands are for camera control, this involves reading the visual imagery commands from the operator station.
   - Read the post-motion commands to determine which sensors should be observed during execution of this command. While some sensors (such as joint limit detectors) are always active, others (such as the gripper contact sensor) are only checked when explicitly requested by the received telemetry.
   - Read and interpret any motion commands for this action. Ground any symbolic symbols in those instructions and convert the motion into a joint space representation.

3. Send the imaging subsystem the current command identifier, current manipulator position, and desired image fragment (extracted from the pre-motion commands).

4. Send the joint space motion command and any required sensory readings to the manipulation subsystem.

5. Now, while the motion is in progress, record joint and sensory data from the manipulation subsystem and update the current joint positions for the imaging subsystem.
6. The motion will continue until any observed sensor changes value or until the motion reaches its natural conclusion (whichever comes first).
7. Once the manipulation subsystem signals that the motion has terminated, parse the command a second time:
8. Phase 2 parser:
   - Step through any real or implied post-motion commands. Evaluate each test condition to determine whether or not an error has occurred during execution and whether or not it is necessary to skip any subsequent commands.
   - If an error has occurred, then:
     - Advise the imaging subsystem to gather and transmit additional imagery to better characterize the error.
     - Use a lookup table to determine a preliminary error diagnosis.
     - Transmit an error signal to the master station.
     - Follow that signal by a detailed history of recorded joint and sensory data for the preceding few seconds of operation.
     - Await the arrival of new post-error instructions from the operator station.
   - If no error has occurred, then evaluate and store any newly-defined symbols before continuing to parse the remaining post-motion commands.
   - Determine which command should next be executed (remember that the post-motion commands may instruct the interpreter to skip forward n commands).

As the interpreter executes each received command, it transmits reply messages, describing the robot's state and any detected errors.

A.4 Slave-to-master communication

Reply messages are sent from the remote slave site back to the operator station. The primary purpose of these messages is to advise the operator when an error has occurred during task execution and to provide him or her with sufficient information to diagnose and correct the problem. A full definition for the reply language is given in Table A.3.

replyStream	$\rightarrow$	*stateReply replies*
replies	$\rightarrow$	*reply replies*
reply	$\rightarrow$	*environmentReply*
	$\vert$	*errorReply priorStates*
priorStates	$\rightarrow$	*stateReply priorStates*
	$\vert$	ϵ
environmentReply	$\rightarrow$	**E** *stateMessage nl*
errorReply	$\rightarrow$	**R** *stateMessage integer float string nl*
stateReply	$\rightarrow$	**S** *stateMessage nl*
stateMessage	$\rightarrow$	*integer float float float float float float float*

TABLE A.3. Language definition for slave to master communication.

A.4.1 The state message

stateMessage $\rightarrow$ *integer float float float float float float float*

The state message describes the robot's state at a particular instant in time. It lists:

1. The integer number of the command that was being executed.

2. The floating-point time in seconds since the first reply was transmitted.

3. The floating-point positions of each joint listed in the order j0, j1, j2... with all values in either radians or millimeters.

A.4.2 The reply stream

$$replyStream \rightarrow stateReply\ replies$$

$$stateReply \rightarrow \mathbf{S}\ stateMessage\ nl$$

$$replies \rightarrow reply\ replies$$

The reply stream begins with a state reply describing the initial robot state and then continues with a stream of either environment or error replies.

A.4.3 The environment reply

$$environmentReply \rightarrow \mathbf{E}\ stateMessage\ nl$$

The environment reply signals that an execution environment has been successfully completed and records the state of the remote robot at the instant the command was completed. For example, if the slave successfully executed the command:

```
B 56
M j0=0.1
M j1=0.2
M j3=0.3
M j4=0.4
M j5=0.5
E
```

then the slave would send a reply message:

```
E 56 743.2 0.1 0.2 0.3 0.4 0.5 0.6
```

This indicates that environment 56 was completed after 743.2 seconds and that joint 0 is at position 0.1, joint 2 at position 0.2 ... and joint 6 is at position 0.6. Note that the positions of all joints are returned even though motion was only commanded for the first five.

A.4.4 The error reply

$$
\begin{array}{rcl}
\textit{reply} & \rightarrow & \textit{errorReply priorStates} \\
\\
\textit{priorStates} & \rightarrow & \textit{stateReply priorStates} \\
 & | & \epsilon \\
\\
\textit{errorReply} & \rightarrow & \mathbf{R}\ \textit{stateMessage integer float string nl} \\
\\
\textit{stateReply} & \rightarrow & \mathbf{S}\ \textit{stateMessage nl}
\end{array}
$$

The error reply is essentially a "request for help" message from the slave. It lists the state the slave was in when the error was detected. It also lists:

1. The integer number of the sensor that detected the error.

2. The floating-point value of that sensor.

3. A string of characters describing the error.

For example, if the slave received the command:

```
B 8
M j6=0.5
E
```

but failed to execute it because a torque limit was exceeded for joint 6 (the gripper joint) then it would send an error reply:

```
E 8 43 0.1 0.2 0.3 0.4 0.5 0.7 26 1 Gripper torque limit
```

As the slave moves, it records its state at regular intervals (every, say, 50–100 ms). During normal operation, the slave only sends a single state message at the end of each environment. Thus, only about one tenth of the recorded state information is transmitted. After an error reply has been transmitted, the slave will send a number of state replies. This state information fills in the gaps between the previously transmitted environment replies. The idea is to give the operator a very detailed picture of the last few seconds of motion leading up to the error. These state messages are sent in reverse time order—that way the operator station receives the most recent (and usually most relevant) information first.

A.4.5 Example reply stream

Here is a typical reply sequence from the slave showing that environments 995–999 have been successfully executed, while there was a problem with environment 1000.

```
E 995   123.1   5  8 30 19 34 15
E 996   124.4   7 13 30 27 39 15
E 997   124.9   9 15 30 29 44 15
E 998   125.4  10 17 30 32 50 15
E 999   126.9  10 20 30 40 50 10
R 1000  127.5  10 20 30 40 50  5 1000 1 Gripper contact
S 1000  127.3  10 20 30 40 50  7
S 1000  127.0  10 20 30 40 50  9
⋮  ⋮        ⋮  ⋮  ⋮  ⋮  ⋮  ⋮  ⋮
S 997   124.7  5  7  30 20 34 15
```

A.4.6 Telemetry

The teleprogramming replies from the remote site are communicated to the master station via the same TCP/IP sockets used for command transmission. Once again, all transmitted telemetry is recorded and timestamped. There is no simulated time delay in the returned signal path—the entire round-trip communications delay having been achieved by delaying transmitted commands on their way to the slave site.

A.5 Interpreting slave replies

As each reply message is received, the master station must remember it in order to enable the replay of prior slave motions following an error. The class tree describing the states used to maintain this historical record is shown in Figure A.6.

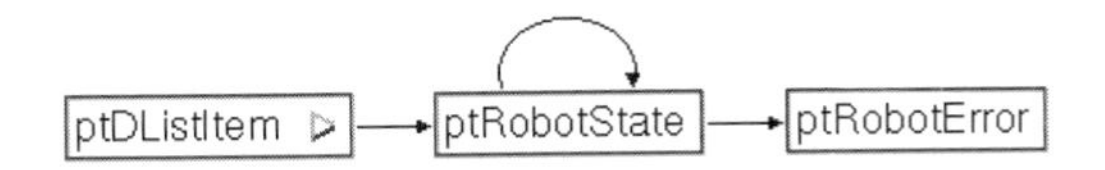

FIGURE A.6: The robot state class tree. The loop-back over the `ptRobotState` indicates a self-referential definition—each `ptRobotState` contains pointers to past and future `ptRobotStates`.

In this case, there are only two types of states. The `ptRobotState` stores the state of the slave at a particular instant in time, as indicated by replies from the slave robot. The `ptRobotErrorState` includes additional information sent by the slave system following detection of an error. Most importantly, it contains the slave's preliminary error diagnosis.

A.6 Maintaining and reviewing the historical record

The operators actions, the generated commands, received replies, and transmitted imagery form multiple parallel data streams. These data streams are stored at the operator station and form a historical record of past activity. After an error, the operator may choose to view simulations of either commanded or measured prior slave motion alongside the corresponding real imagery.

Referring to the stylized example of concurrent data streams shown in Figure A.7, there are six important features:

1. During the initial task execution, the operator performs a sequence of motions that are translated into commands. These are successfully executed by the slave, which generates both symbolic replies and visual imagery. Note that several of the operator's actions were sufficiently similar that they did not require the generation of unique commands. The system indexes the replies and received imagery back into the stream of generated commands.

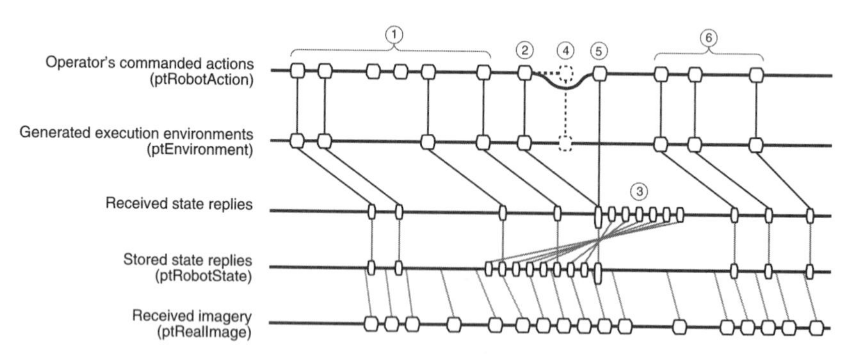

FIGURE A.7: The implementation manipulates multiple parallel data streams (see text for detailed description).

2. In this case the operator performed an action that (unknown to him or her) will fail to execute correctly at the slave site

3. When the slave detects the error, it sends an error reply followed by a detailed history of past slave actions. This detailed history is sent in reverse time order and “fills in the blanks” between the previously received replies. Thus, the stored state replies need not maintain the same ordering as the received state replies.

4. This represents an example of an action performed by the operator after generation of a command that will ultimately fail, but before the master station is aware of that failure. Commands of this type will never be executed by the slave. When the error reply is received, the master station will “wind back time”, undoing the effects of any such actions.

5. This action is inserted into the action stream by the system upon receipt of the error signal. It moves the simulated slave robot to the position of the real robot just after the error was detected.

6. After diagnosing the error, the operator continues with the task, generating new commands for the remote robot.

The operator interface to the historical record is through the error recovery dialog (see Figure A.8). This advises the operator of which command was executing when the error was detected, along with the slave system's preliminary error diagnosis. The operator may then review the historical record of expected slave motion (from the `ptRobotAction` list) or actual recorded slave motion (from the `ptRobotState` list). He or she may request that the system replay the last few seconds of operation or, alternatively, the

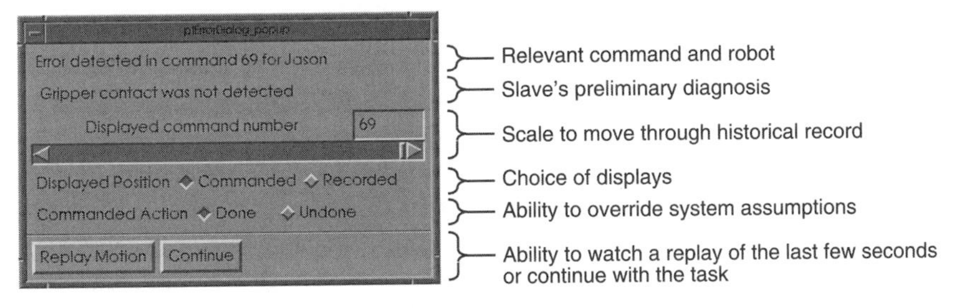

FIGURE A.8: The error dialog box provides the operator interface to the historical record.

operator may manually slide forward and back through the stored record. Once satisfied with a diagnosis, he or she may continue with the task.

Looking back on the implementation, it was the ability to command a robot from a distance which made it possible to give compelling laboratory demonstrations. But it was the ability to continue after an error, which made it possible to build a real-world implementation. Just as in life, one must learn how to fall, and stand back up, before one can learn to run.

References

[1] Steven Abrams, Peter K. Allen, and Konstantinos A. Tarabanis. Dynamic sensor planning. In *Proceedings DARPA Image Understanding Workshop*, Pittsburgh, April 1993.

[2] Alfred V. Aho, John E. Hopcroft, and Jeffrey D. Ullman. *The Design and Analysis of Computer Algorithms.* Addison-Wesley, 1974.

[3] Robert J. Anderson. Teleoperation with virtual force feedback. In *Third International Symposium on Robotics Research*, pages 125–131, October 1993.

[4] Russell L. Andersson. Computer architectures for robot control: A comparison and a new processor delivering 20 real Mflops. In *Proceedings IEEE Conference on Robotics and Automation*, pages 1162–1167, Scottsdale, Arizona, May 1989.

[5] Margo K. Apostolos, Haya Zak, Hari Das, and Paul Schenker. Multisensory feedback in advanced teleoperations: Benefits of auditory clues. In *Sensor Fusion V, SPIE, vol. 1828*, pages 98–105, November 1992.

[6] Autodesk Inc. *AutoCAD Reference Manual*, 1989.

[7] A.K. Bejczy. Sensors, controls and man-machine interface for advanced teleoperation. *Science*, 208:1327–1335, June 1980.

[8] Anatal K. Bejczy, Steven Venema, and Won S. Kim. Role of computer graphics in space telerobotics: Preview and predictive displays. In *Cooperative Intelligent Robotics in Space*, pages 365–377. Proceedings of the SPIE, vol. 1387, November 1990.

[9] Eric Allan Bier and Maureen C. Stone. Snap-dragging. *SIGGRAPH Proceedings*, 20(4):233–240, August 1986.

[10] Edward G. Britton, James G. Lipscomb, and Michael E. Pique. Making nested rotations convenient for the user. In *ACM SIGRAPH*, pages 222–227, 1978.

[11] Frederick P. Brooks, Ming Ouh-Uoung, James J. Batter, and P. Jerome Kilpatrick. Project GROPE – haptic displays for scientific visualization. *ACM Computer Graphics*, 24(4):177–185, 1990.

[12] Thurston Brooks, Ilhan Ince, and Greg Lee. Operator vision aids for space teleoperation assembly and servicing. In *Fifth Annual Space Operations, Applications and Research Symposium*, 1991.

[13] Guy Bruno and Matthew K. Morgenthaler. Real time proximity cues for teleoperation using model based force reflection. In *Cooperative Intelligent Robotics in Space II*, pages 346–355. SPIE, vol. 1612, November 1991.

[14] Consortium TELEROBOT, Genoa, Italy. *MASCOT: Technical Description*, 1992.

[15] Lynn Conway, Richard Volz, and Michael Walker. Tele-autonomous systems: Methods and architectures for intermingling autonomous and telerobotic technology. In *Proceedings of the IEEE International Conference on Robotics and Automation*, pages 1121–1130. IEEE, March 1987.

[16] Lynn Conway, Richard A. Volz, and Michael W. Walker. Teleautonomous systems: Projecting and coordinating intelligent action at a distance. *IEEE Transactions on Robotics and Automation*, 6(2):146–158, April 1990.

[17] James O. Coplien. *Advanced C++, Programming Styles and Idioms.* Addison-Wesley, 1992.

[18] John J. Craig. *Introduction to Robotics: Mechanics and Control.* Addison-Wesley, second edition, 1989.

[19] Allen Cypher and Daniel C. Halbert (ed). *Watch what I do : Programming by demonstration.* M.I.T. Press, 1993.

[20] Department of the Navy, Headquarters Naval Material Command. *Physics of Sound in the Sea*, 1969.

[21] Robin M. Dunbar et al. Autonomous underwater vehicle communications. In *ROV '90*, pages 270–278. The Marine Technology Society, June 1990.

[22] E.R. Ellis. Nature and origins of virtual environments: A bibliographical essay. *Computing Systems in Engineering*, 2(4):321–347, 1991.

[23] R.E. Ellis, O.M. Ismaeli, and M.G. Lipsett. Design and evaluation of a high-performance prototype planar haptic interface. In *Advances in Robotics Mechatronics and Haptic Interfaces, ASME Winter Annual Meeting, DSC Vol 49*, pages 55–64, November 1993.

[24] M.J. Feldman and W.R. Hamel. The advancement of remote systems technology: Past perspectives and future plans. In *Remote Handling in Nuclear Facilities*, pages 50–69. OECD Nuclear Energy Agency, 1984.

[25] William R. Ferrell. Remote manipulation with transmission delay. *IEEE Transactions on Human Factors in Electronics*, HFE-6(1):24–32, 1965.

[26] James D. Foley, Andries van Dam, Stephen K. Feiner, and John F. Hughes. *Computer Graphics: Principles and Practice.* Addison-Wesley, second edition, 1990.

[27] Toshio Fukuda and Kazuhiro Kosuge. Advanced telerobotic systems: Single-master multi-slave manipulator system and cellular robotic system. *Teleoperation: Numerical Simulation and Experimental Validation. M.C. Becquet (ed)*, pages 195–208, 1992.

[28] Janez Funda. *Teleprogramming: Towards delay-invariant remote manipulation.* PhD thesis, The University of Pennsylvania, 1991.

[29] Janez Funda, 1994. Presentation on surgical robotics at IBM given at the GRASP laboratory of the University of Pennsylvania.

[30] Janez Funda, Thomas S. Lindsay, and Richard P. Paul. Teleprogramming: Toward delay-invariant remote manipulation. *Presence*, 1(1):29–44, Winter 1992.

[31] S.V. Gray, J.R. Wilson, and C.S. Syan. Human control of robot motion: Orientation, percention and compatability. In Mansour Rahimi and Waldemar Karwowski, editors, *Human-Robot Interaction*, chapter 3, pages 48–64. Taylor and Francis, 1992.

[32] Gregory D. Hager, Gerhard Grunwald, and Gerd Hirzinger. Feature-based visual servoing and its application to telerobotics. In *IROS*, pages 164–171, 1994.

[33] Samad H. Hayati. Position and force control of coordinated multiple arms. *IEEE Transactions on Aerospace and Electronic Systems*, 24(5):584–590, September 1988.

[34] G Hirzinger. ROTEX the first robot in space. In *International Conference on Advanced Robotics*, pages 9–33, November 1993.

[35] David F. Hoag and Vinay K. Ingle. Underwater image compression using the wavelet transform. In *Oceans*, pages 533–537, Brest, France, September 1994.

[36] R.L. Hollis and S.E. Salcudean. Lorentz levitation technology: A new approach to fine motion robotics, teleoperation, haptic interfaces and vibration isolation. In *Preprints, Sixth International Symposium on Robotics Research*, October 1993.

[37] John E. Hopcroft and Jeffrey D. Ullman. *Introduction to Automota Theory Languages and Computation.* Addison-Wesley, 1979.

[38] S. Hyati, T. Lee, K. Tso, P. Backes, and J. Lloyd. A testbed for unified teleoperated-autonomous dual-arm robotic system. In *IEEE Robotics and Automation Conferernce*, pages 1090–1095, 1990.

[39] K. Ikeuchi. Assembly plan from observation. In Takedo Kanade and Richard Paul, editors, *Preprints of the Sixth International Symposium on Robotics Research*, October 1993.

[40] R.F. Jackson. Remote handling—u.k. overview. In *Remote Handling in Nuclear Facilities*, pages 22–49. OECD Nuclear Energy Agency, 1984.

[41] Timothy E. Johnson. Sketchpad III—a computer program for drawing in three dimensions. In *Proceedings—Spring Joint Computer Conference (AFIPS)*, pages 347–353, 1963.

[42] Won S. Kim and Anatal K. Bejczy. Graphics displays for operator aid in telemanipulation. In *IEEE International Conference on Systems Man and Cybernetics*, Charlottesville, VA, October 1991.

[43] Won S. Kim and Lawrence W. Stark. Cooperative control of visual displays for telemanipulation. In *Proceedings of the IEEE International Conference on Robotics and Automation*, pages 1327–1332. IEEE, May 1989.

[44] Tetsuo Kotoku, Kazuo Tanie, and Akio Fujikawa. Force-reflecting bilateral master-slave teleoperation system in virtual environment. In *International Symposium on Artificial Intelligence, Robotics and Automation in Space*, 1990.

[45] Angela Lai. Operator-assisted model-based vision. Master's thesis, The University of Pennsylvania, 1994.

[46] M.H.E. Larcombe and J.R. Halsall. *Robotics in Nuclear Engineering.* Graham and Trotman Ltd., 1984.

[47] Thomas S. Lindsay. *Teleprogramming: Remote Site Robot Task Execution.* PhD thesis, The University of Pennsylvania, 1992.

[48] Tomás Lozano-Pérez. Robot programming. *Proceedings of the IEEE*, 71(7):821–841, July 1988.

[49] Kazuo Machida, Yoshitsugu Toda, and Toshiaki Iwata. Graphic-simulator-augmented teleoperation system for space applications. *Journal of Spacecraft and Rockets*, 27(1):64–69, 1990.

[50] R.L. Marks, H.H.Wang, S.M.Rock, and M.J.Lee. Automatic visual stationkeeping of an underwater robot. In *Oceans*, pages II–137–II–142, September 1994.

[51] Michael J. Massimino and Thomas B. Sheridan. Variable force and visual feedback effects on teleoperator man/machine performance. In *Proceedings of the NASA Conference on Space Telerobotics*, pages 89–98, January 1989.

[52] Douglas A. McCaffee. Diverse applications of advanced man-telerobotic interfaces. In *Technology 2000*, pages 350–360, March 1991.

[53] Gerard T. McKee and Paul S. Schenker. Visual acts for remote viewing during teleoperation. In *IEEE Robotics and Automation Conference, Nagoya, Japan*, 1995.

[54] Pat P. Miller. *Script Supervising and Film Continuity.* Focal Press, 1986.

[55] Paul Millman, Michael Stanley, Paul Grafing, and J. Edward Colgate. A system for the implementation and kinesthetic display of virtual environments. In *SPIE, vol. 1833 Telemanipulator Technology*, pages 49–56, Boston, MA, November 1992.

[56] Brad A. Myers. Demonstrational interfaces: A step beyond direct manipulation. *IEEE Computer*, pages 61–73, August 1992.

[57] Ming Ouh-Young, David V. Beard, and Frederick P. Brooks, Jr. Force display performs better than visual display in a simple 6-D docking task. In *Proceedings of the IEEE International Conference on Robotics and Automation*, pages 1462–1466. IEEE, May 1989.

[58] Eimei Oyama, Naoki Tsunemoto, Susumu Tachi, and Yasuyuki Inoue. Experimental study on remote manipulation using virtual reality. *Presence*, 2(2):112–124, Spring 1993.

[59] Richard P. Paul. *Robot manipulators : Mathematics, programming, and control.* M.I.T. Press, 1981.

[60] Richard P. Paul, Bruce Shimano, and Gordon E. Mayer. Kinematic control equations for simple manipulators. *IEEE Transactions on Systems Man and Cybernetics*, 11(6):449–455, 1981.

[61] Edward Pincus. *Guide to Filmmaking.* Signet, 1969.

[62] Majid Rabbani and Paul W. Jones. *Digital Image Compression Techniques.* SPIE Optical Engineering Press, 1991.

[63] Daryl Rasmussen. A natural visual interface for precision telerobot control. In *SPIE, vol. 1833 Telemanipulator Technology*, pages 170–179, Boston, MA, November 1992.

[64] Jr. Robert H. Sturges. Reliability and safety in teleoperation. In James H. Graham, editor, *Safety, Reliability and Human Factors in Robotic Systems*, chapter 5, pages 83–115. Van Norstrand Reinhold, 1991.

[65] Roger Y. Tsai. A versatile camera calibration technique for high-accuracy 3D machine vision metrology using off-the-shelf TV cameras and lensses. *IEEE Journal of Robotics and Automation*, RA-3(4):323–344, August 1987.

[66] Louis B. Rosenberg. The use of virtual fixtures as perceptual overlays to enhance operator performance in remote environments. Technical Report AL-TR-1992-XXXX, USAF Armstrong Laboratory, Wright-Patterson AFB OH, September 1992. (in publication).

[67] A. Rovetta, F. Cosmi, L. Molinari Tosatti, and L. Termite. Evaluation of human control in telerobotics by means of EMG. In *IROS*, pages 268–272, 1994.

[68] Craig Sayers and Richard Paul. Synthetic fixturing. In *Advances in Robotics, Mechatronics and Haptic Interfaces (DSC–Vol.49) ASME Winter Annual Meeting, New Orleans, USA*, pages 37–46, November 28–December 3 1993.

[69] Craig Sayers and Richard Paul. An operator interface for teleprogramming employing synthetic fixtures. *Presence*, 3(4):309–320, 1994.

[70] Craig Sayers, Richard Paul, and Max Mintz. Operator interaction and teleprogramming for subsea manipulation. In *Fourth IARP Workshop on Underwater Robotics, Genova, Italy*, November 1992.

[71] Craig Sayers, Richard Paul, Louis Whitcomb, and Dana Yoerger. Subsea teleoperation from untethered vehicles via acoustic communication. *IEEE Journal of Oceanographic Engineering*, 1995. (in preparation).

[72] Craig Sayers, Matthew Stein, Angela Lai, and Richard Paul. Teleprogramming to perform sophisticated underwater manipulative tasks using acoustic communications. In *OCEANS '94, Special Session on Deep Automated Unmanned Vehicles*, pages 168–173, Brest, France, September 1994.

[73] Brian Schmult and Robert Jebens. Application areas for a force-feedback joystick. In *Advances in Robotics Mechatronics and Haptic Interfaces, ASME Winter Annual Meeting, DSC Vol 49*, pages 65–72, November 1993.

[74] Stanley A. Schneider and Robert H. Cannon Jr. Experimental object-level strategic control with cooperating manipulators. *International Journal of Robotics Research*, 12(4):338–350, August 1993.

[75] SensAble Devices Inc., USA. *The PHANToM Force-Reflecting Haptic Interface.*

[76] T.B. Sheridan. Telerobotics. *Automatica*, 25(4):487–507, July 1989.

[77] Thomas Sheridan. *Telerobotics, Automation, and Human Supervisory Control.* M.I.T. Press, Cambridge, Mass., 1992.

[78] Thomas Sheridan. Space teleoperation through time delay:Review and prognosis. *IEEE Transactions on Robotics and Automation*, 9(5):592–606, October 1993.

[79] Matthew R. Stein and Richard P. Paul. Behavior based control in time delayed teleoperation. In *Sixth International Conference on Advanced Robotics - '93 ICAR*, pages 223–228, 1993.

[80] W. Richard Stevens. *Unix Network Programming.* Prentice-Hall, 1990.

[81] Melica Stojanovic, Zoran Zvonar, Josko A. Catipovic, and John J. Proakis. Spatial processing of broadband underwater acoustic communication signals in the presence of co-channel interference. In *Oceans*, pages 286–291, Brest, France, September 1994.

[82] Bjarne Stroustrup. *The C++ Programming Language.* Addison-Wesley, second edition, 1991.

[83] Ivan E. Sutherland. Sketchpad—a man-machine graphical communication system. In *Proceedings - Spring Joint Computer Conference (AFIPS)*, pages 328–346, 1963.

[84] Ivan E. Sutherland. The ultimate display. In *Proceedings of the IFIP Congress*, pages 506–508, 1965.

[85] Kazuo Tanie and Tetsuo Kotoku. Force display algorithms. In *Lecture Notes for Workshop on Force Display in Virtual Environments and its Application to Robotic Teleoperation*, pages 60–78. IEEE International Conference on Robotics and Automation, May 1993.

[86] Konstantinos Tarabanis and Roger Y. Tsai. Sensor planning for robotic vision: A review. In Craig Khatib and Lozano-Pérez, editors, *Robotics Review 2*, pages 113–136, Cambridge, 1992. M.I.T. Press.

[87] T.J. Tarn, A.K.Bejczy, and X. Yun. New nonlinear control algorithms for multiple manipulators. *IEEE Transactions on Aerospace and Electronic Systems*, 24(5):584–590, September 1988.

[88] Russel H. Taylor. An overview of computer assisted surgery at IBM T.J. Watson Research Center. Technical Report RC 19166 (83465) 9/16/93, IBM, 1993.

[89] Stanley Unruh, Terry Faddis, and Bill Barr. Possition assist shared control of a force reflecting telerobot. In *SPIE, vol. 1833 Telemanipulator Technology*, pages 113–121, Boston, MA, November 1992.

[90] Jean Vertut and Philippe Coiffet. *Robot Technology*, volume 3A Teleoperations and Robotics: Evolution and Development. Prentice-Hall, 1986.

[91] Jean Vertut and Philippe Coiffet. *Robot Technology*, volume 3B Teleoperations and Robotics: Applications and Technology. Prentice-Hall, 1986.

[92] Yujin Wakita, Shigeoki Hirai, and Toshiyuki Kino. Automatic camera-work control for intelligent monitoring of telerobotic tasks. In *IEEE/RSJ International Conference on Robotics and Systems*, pages 1130–1135, July 1992.

[93] Leonard R. Wanger, James A. Ferwerda, and Donald P. Greenberg. Perceiving spatial relations in computer-generated images. *IEEE Computer Graphics and Applications*, 12(2):44–58, May 1992.

[94] Louis L. Whitcomb and Dana R. Yoerger. A new distributed real-time control system for the JASON underwater robot. In *IROS*, pages 368–374, Yokohama, Japan, July 1993.

[95] Matthias M. Wloka. Lag in multiprocessor virtual reality. *Presence*, 4(1):50–63, 1995.

[96] Stephen Wolfram. *Mathematica: A system for doing mathematics by computer.* Addison-Wesley, second edition, 1991.

[97] Dana R. Yoerger. *Supervisory Control of Underwater Telemanipulators:Design and Experiment.* PhD thesis, M.I.T., 1982.

[98] Tsuneo Yoshikawa. *Foundations of Robotics: Analysis and Control.* M.I.T. Press, 1990.

[99] Herbert Zettl. *Sight Sound Motion: Applied Media Aesthetics.* Wadsworth Publishing, second edition, 1990.

[100] Y.F. Zheng and J.Y.S. Luh. Joint torques for control of two coordinated moving robots. In *Proceedings of the IEEE International Conference on Robotics and Automation*, pages 1375–1380. IEEE, 1986.

Index